Springer Tracts on Transportation and Traffic

Volume 6

Series editor

Roger P. Roess, New York, USA

About this Series

The book series "Springer Tracts on Transportation and Traffic" (STTT) publishes current and historical insights and new developments in the fields of Transportation and Traffic research. The intent is to cover all the technical contents, applications, and multidisciplinary aspects of Transportation and Traffic, as well as the methodologies behind them. The objective of the book series is to publish monographs, handbooks, selected contributions from specialized conferences and workshops, and textbooks, rapidly and informally but with a high quality. The STTT book series is intended to cover both the state-of-the-art and recent developments, hence leading to deeper insight and understanding in Transportation and Traffic Engineering. The series provides valuable references for researchers, engineering practitioners, graduate students and communicates new findings to a large interdisciplinary audience.

More information about this series at http://www.springer.com/series/11059

Tomaž Tollazzi

Alternative Types of Roundabouts

An Informational Guide

Springer

Tomaž Tollazzi
Department of Traffic Management
Faculty of Civil Engineering
University of Maribor
Maribor
Slovenia

ISSN 2194-8119 ISSN 2194-8127 (electronic)
ISBN 978-3-319-37864-0 ISBN 978-3-319-09084-9 (eBook)
DOI 10.1007/978-3-319-09084-9

Springer Cham Heidelberg New York Dordrecht London

Printed on acid-free paper

Springer is part of Springer Science+Business Media (www.springer.com)

To Michael E. Brown

Preface

Over recent decades roundabouts have become increasingly used when building new at-grade intersections or up-grading junctions all over the world, and when rebuilding existing intersections. However, control of traffic flows at at-grade intersections and up-grade junctions by roundabouts creates unique design problems. The history of researching roundabouts shows that "what is going on" is not always obvious. The first theories and studies were influenced by the existing urban road layouts but changes in vehicles' constructions, dimensions, and speeds also had a strong impact. Today, after many years of experience, there have been different ideas about the "ideal roundabout", with little consensus about the crucial effects of the rules on how to negotiate an intersection.

Today, modern roundabouts exist in all European countries and elsewhere in the world. We can now say that modern roundabouts are a world phenomenon. In Europe, there are no uniform guidelines for the geometric designing of roundabouts, which is understandable because the situation in one country is very different from another. A certain solution which would be safe from the traffic safety point of view in one country could be very dangerous in another. Consequently, most countries have their own guidelines for the geometric designing of roundabouts which are, as far as possible, adapted to real circumstances (local customs, habits, traffic culture…) within these countries and are therefore the most acceptable within their surroundings.

Roundabouts in different countries also differ in their dimensions and designs, the reasons for this being the different maximum dimensions of motor vehicles (mostly heavy vehicles), and human behavior.

In the cases of roundabouts, there is not "only one truth". Therefore, each country needs to "walk its own path", although this is maybe the slowest and the more difficult, it is also the safest way. Verbatim, the copying of foreign results could be dangerous and could lead to effects that are completely the opposite than expected.

It needs to be stressed that the roundabout intersection has been "at the development phase" since 1902, and this development is still in progress. One of the results of this progress is the several types of roundabouts in worldwide usage today, called the "alternative types of roundabouts". Some of them are already in frequent use all over the world, some of them are recent and have only been implemented within certain countries, and some of them are still at the development phase. Alternative types of roundabouts typically differ from "standard" one- or two-lane roundabouts in one or more design elements, as their purposes for implementation are also specific. The main reasons for their implementation are the particular disadvantages of "standard" one- or two-lane roundabouts regarding actual specific circumstances. Usually, these disadvantages are highlighted by low levels of traffic safety or capacities.

Therefore, it was decided that it would be useful to collate in one book some of the alternative types that are already in frequent use today in some countries and those that are "still coming".

The content of this book is as follows:

Chapter 1 deals with the origins of roundabouts, squares, traffic centers, traffic islands, and their early developments.

Chapter 2 deals with the developments of different roundabout types. The chapter starts with the first trends and then to non-circular islands and larger roundabouts. The remaining part of the chapter is dedicated to the period of intensive experimentation with new layouts, and to some of the research resulting in the implementation in real life of different types of roundabouts.

Chapter 3 presents the modern layout designs of roundabouts, the criteria for the acceptability of roundabouts, as required or recommended in some countries, their geometric design features, the effects of layout design elements on traffic safety, and some European and non-European countries experiences with traffic safety on roundabouts.

Chapter 4 deals with recent alternative types of roundabouts. The chapter starts with their definitions and design characteristics. The remaining part of the chapter is dedicated to some of today's alternative types of roundabouts.

Chapter 5 presents some of the alternative types of roundabouts at development phases, basic ideas and their characteristics, design elements, and capacities.

Chapter 6 is dedicated to the general criteria for calculating the capacities of alternative types of roundabouts. A short state-of-the-art is presented, three basic models, brief information on each of them, with emphasis on the capacity calculations of alternative types of roundabouts at the development phases.

Chapter 7 deals with non-motorized participants on recent alternative types of roundabouts and roundabouts at development phases.

Chapter 8 presents possible ways of roundabouts' developments and some directions for future research. This chapter has a more general scope because the situation can differ from country to country, depending on their experiences with existing standard roundabouts.

The book includes several schemes, drawings, and figures that help the reader to better understand the material.

Maribor, Slovenia Tomaž Tollazzi

Acknowledgments

My interest in roundabouts started in 1990 when, as a young civil engineer, I met Michael E. Brown in London. Mike was my first mentor in the study of roundabouts. Thank you Mike!

Most of this book is the result of my 20 years' study of roundabouts, many visits to several countries using roundabouts (the UK, The Netherlands, Germany, Spain, Portugal, France, Switzerland, Denmark, Belgium, Malta, Mexico, the USA...), and cooperation with professors on their faculties, responsible persons for their guidelines regarding roundabouts, and their designers, in these countries.

Thanks to those who helped me in the beginning to be become acquainted with the secrets and basics of roundabouts, as well as those from whom I have drawn on their knowledge over the later period.

I would also like to thank all those who contributed to the producing of this book in any way. Special thanks to Prof. Raffaele Mauro, Prof. Werner Brilon, Prof. Hideki Nakamura, L.G.H. Fortuijn, Marco Guerrieri, Clive Sawers, Rijkswaterstaat Holland, Peter Wijnands, Zoran Kenjić, Janet Barlow, Tedi Zgrablić, Yok Hoe Yap, Jure Bergoč, Drago Bole, Darko Gorenak, Martin Smělý, Goran Jovanović, Jiří and Tomas Apeltauer, Tiziana Campisi, Bill Baranowski, Saulius Vingrys, Tim Murphy, Keith Boddy, Nova Scotia Transportation and Infrastructure Renewal, Ivica Barišić, Jovan Hristoski, and Dirk de Baan.

Special thanks also to Marko Renčelj, Sašo Turnšek, and Sanja Vodnik, my colleagues from the University of Maribor, for their help.

Thanks also to the Slovenian Ministry of Infrastructure and Spatial Planning, the Slovenian Roads Agency for supporting many of my research projects on roundabouts.

Thanks also to Prof. Branko Bedenik and George Yeoman for technical and grammar proofreading.

Finally, I wish to thank Norah Jones and Eric Bibb for their company throughout the writing of this book.

Contents

Chapter 1
Origins of Roundabouts

1.1 Introduction

In a comprehensive review of roundabouts it could firstly be beneficial to look at their origins. The conventions or rules of usage are as important as the layout. The history of roundabouts shows that these conventions must be conveyed by clear signs and warnings, and by the indications implicit within the layout. If this is done well then drivers will respond logically including moderating their speed.

Experience indicates that drivers will readily learn to exploit the junction, often to an extent exceeding previous expectance of capacity. The layout will also be safe if the arrangement induces the necessary decisions and maneuvers to be made at a speed within the average human capability [1].

At the beginning, and for better understanding, it is necessary to define the meanings of certain terms, like the differences between "squares", "traffic islands", "modern roundabouts" and "alternative types of roundabouts".

At the beginning, the square was intended only for pedestrians and horse-drawn vehicles, and it was much later that cyclists and motor vehicles were included. An elevated platform, as a rule, did not exist on squares or there was not a circular island.

Later it became necessary to separate pedestrians from motorized traffic. Thus, at the outer edges of the square, the resulting surface, which was intended only for motor vehicles and a resting place, became an elevated platform intended for pedestrians only. These were the first traffic circles. At that time there was no need for traffic rules since there was a low traffic volume and low speeds. Later, the first traffic rules were introduced, including that vehicles at the entrance to the traffic circle had the right-of-way.

In the mid-twentieth century a "modern roundabout" was developed in the United Kingdom by their Transport Research Laboratory. On a modern roundabout, priority is given to the circulating flow, and the central island provides a visual barrier across the intersection to the drivers entering it. The main point of the term "modern

T. Tollazzi, *Alternative Types of Roundabouts*, Springer Tracts on Transportation and Traffic 6, DOI 10.1007/978-3-319-09084-9_1

roundabouts" is to emphasize the distinction regarding the older "traffic circle" junction types which had different design characteristics and rules of operation. Older "traffic circles" or "rotaries" were typically larger, operated at higher speeds, and often gave priority to entering traffic. The main characteristic of a "modern roundabout" is the deflection on entry, and the speed around the central island is about 25–40 km/h. Analysis of literature shows that nowadays "modern roundabouts" exist in all European countries, as well as in more than 60 countries elsewhere in world.

Today in some European countries, instead of the term "modern roundabout" the term "standard roundabout" is used which is understandable since the "moderns" are already long-time "standards", especially in countries in which "modern roundabouts" have been implemented over several decades. The point of the term "standard roundabouts" (standard layout means simple one- or two-lane roundabouts) is also to emphasize the distinction from the "alternative types of roundabouts".

There are several types of roundabouts in worldwide usage today, called the "alternative types of roundabouts". Some of them are already very old, and some of them are still quite new. Some of them have only been implemented in certain countries but some of them are in frequent usage all over the world. They typically differ from the "standard" one- or two-lane roundabouts in one or more design elements, as their purposes for implementation are also specific. The main reasons for their implementation are the particular disadvantages of "standard" one- or two-lane roundabouts regarding actual specific circumstances. Usually, these disadvantages are highlighted by low-levels of traffic safety or capacities.

1.2 Town Squares and Traffic Centers

With the development of independent principalities, kingdoms and finally, absolutism in Western Europe, the conditions for the realizations of "ideal towns" were created. The central spot of a sovereign's capital was devoted to the monarch's palace, residence or the summer mansion. Rulers all over Europe were creating new capitals with dazzling squares and parks. The wealthiest absolute monarchies also built the largest baroque and classicistic urban compositions. The leading country in this was France (Fig. 1.1), than copied by the tsarist Russia, Germany, England, Italy etc. Symmetric axes compositions with diagonal, star-shaped squares (Fig. 1.2) and other axial features, appearing in the *points de vue* of palaces, obelisks, fountains… became a regular repertoire of baroque urbanism [2].

In the context of that time, a "roundabout" was created in France, originating from *rond-point* or *point circulaire* (circular point). Namely, in large French royal parks, strollers and often also horsemen might have got lost; therefore, each connection of major diagonals in parks and forests designed in such a "French style" was an appropriate place for lost strollers and horsemen to meet other people from their escort again. Moreover, *rond-points* were ideal places for setting up monuments, fountains and obelisks, which increased the possibility of orientation, at the beginning in the parks, and later also in the cities [3].

Fig. 1.1 Paris, view of the park Jardin des Tuileries—old postcard, 1899

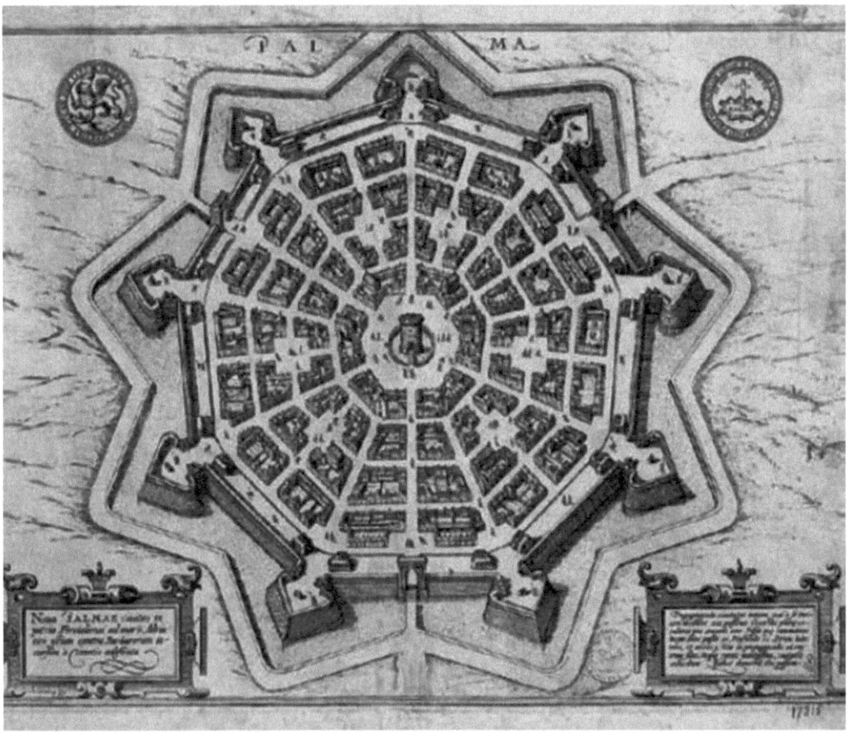

Fig. 1.2 City of Palmanova, Italy; established 1593 [4]

At the end of the 18th century, the symbols of absolute power, expressed by the central position of the mansion, palace, etc. were joined by the entirely new urban demands of functional and technological character. The transport of animal-drawn vehicles (Fig. 1.3) in the big cities of the time (Paris, London…) increased to the point where wide, level penetrations of the roads were required through the spiral medieval structures.

The greatest breakthrough in factors affecting urbanism came at the end of 18th and in the first half of the 19th century, commonly referred to as the industrial revolution. The most important invention that affected the development of industry was the steam engine, which enabled the creation of large factories no longer dependent on workforces from the East.

In the 19th century, the city population was growing quickly. In particular, the great metropolises—ports (Fig. 1.4) and industrial and trade centers, were growing rapidly. London, for example, had 450,000 inhabitants in 1660, while the number increased to 3,000,000 in 1860 and to 9,000,000 in 1960 [1].

At the end of the 19th century, urban regulation was mostly limited to the leveling and widening of roads, the formations of rings and city parks, the determinations of building lines and the heights of the buildings. Despite the above-mentioned negative aspects of urbanization during the industrial revolution, the 19th century also represented the beginnings of modernizing cities, the regulation of water supplies, sewage systems, gas, electricity, public city transport, lighting, and communal services [2].

Fig. 1.3 Town square in Granada—old postcard, 1900

Fig. 1.4 Marseille, Republic Street—old postcard, 1905

Places and sites of circular shape have been an instrument of city and town planning since the middle ages and especially from the renaissance to the 19th century. Circular places were used at the convergent points of road systems (Fig. 1.5). This became a particular feature in the reconstruction of Paris, Vienna, and other cities (Fig. 1.6) with the establishments of imposing places at the intersections of wide boulevards and other radiating roads [1].

Fig. 1.5 Paris—old postcard, 1910

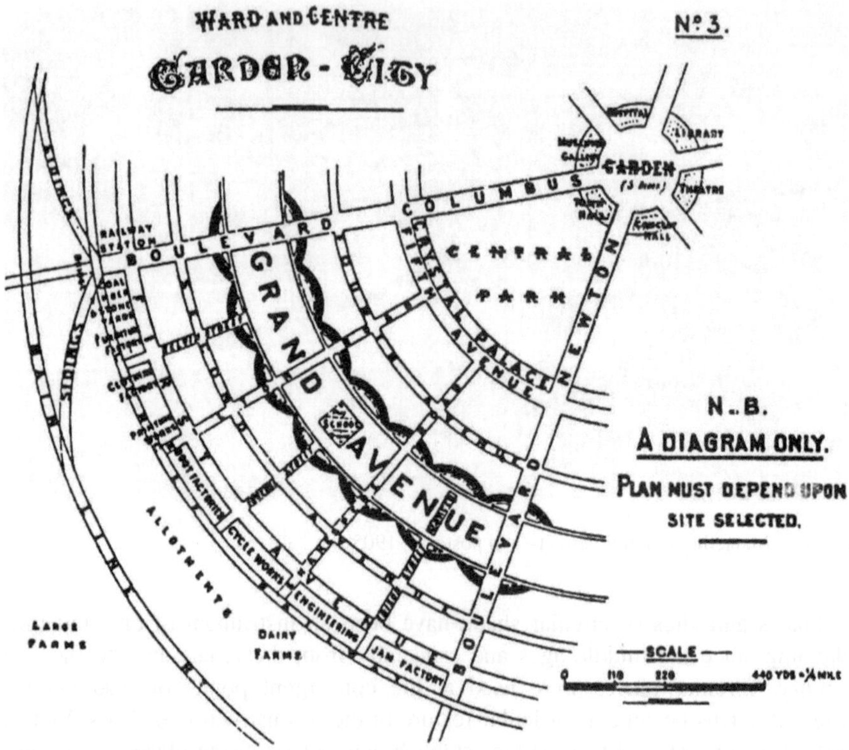

Fig. 1.6 "A garden city", 1902 [1]

There were a number of these centers in Haussmann's grand design for Paris, where traffic could be induced to circulate around a central monument. These points regarding concentrated traffic were to prove a recipe for traffic congestion, even with the horse-drawn vehicles of the time, and providing efficient control of conflicting traffic was a serious problem for the authorities [2].

1.3 First Concepts and the Early Development

The idea of the gyratory operation dates from at least 1903, when Eugène Hénard proposed the circular course of traffic as a solution to the problem of dense traffic in the centers of big cities. This required all the traffic to circulate in one direction (Fig. 1.7). The first practical application of a gyratory system appears to have been the "Columbus Circle", installed by William Phelps Eno in New York in 1905 [1].

An even earlier implementation of Hénard's "giratoire-boulevard" principle is indicated by the proposals for the reconstruction of the main streets in Lisbon. Frederico Ressano Garcia, Hénard's contemporary, proposed such a type of

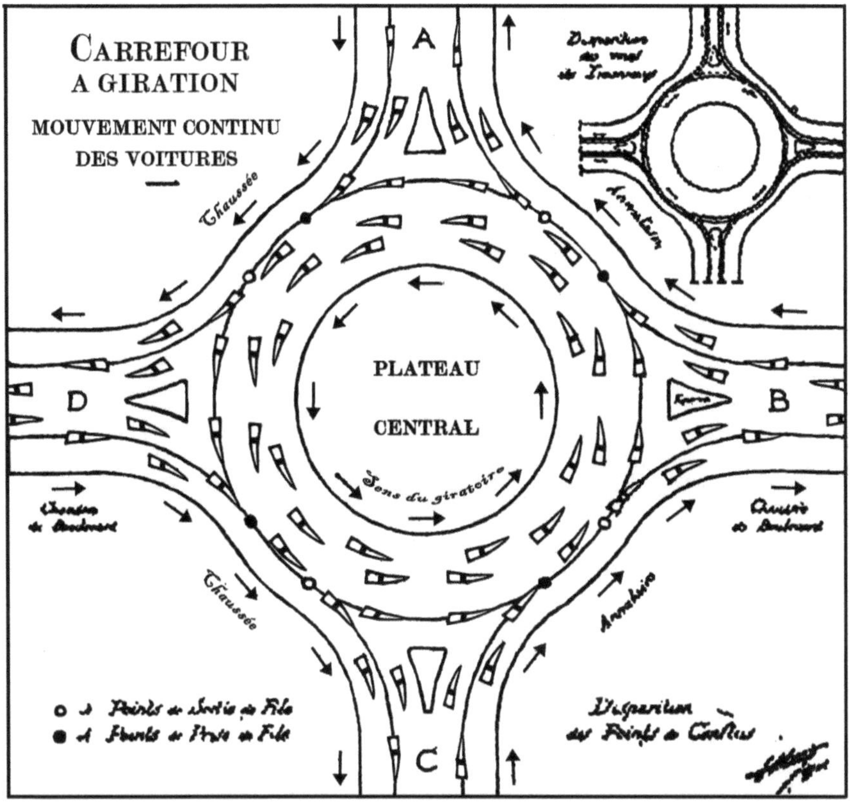

Fig. 1.7 "Gyratory system", suggested by Hénard [1]

intersection in 1877, at the l`École Impériale des Ponts et Chaussées in Paris. This system was not inaugurated in Paris at the Place de l'Etole, where twelve radial roads met, until 1907 (Fig. 1.8). It was also introduced in the Place de la Nation, intersecting ten roads [1].

Around the same time (1905), there was a traffic revolution with rapidly increasing numbers of motor vehicles. Both big countries, United Kingdom and USA, were under the influence of the French post-Napoleon road systems at the time. Therefore, the idea of improving the roads corresponding to the large increase in the motorized traffic worked well in the United Kingdom. At that time Parker and Unwin were developing Ebenezer Howard's First Garden City at Letchworth in Herdfordshire. The essence of Howard's method of traffic management was roads leading from the center of the city (central hub) outwards (radial street system). During a visit to Paris in 1908, Parker was so impressed with the Place de l`Étoile, that on his return a six-arm intersection planned for Letchworth was designed as a gyratory. This first roundabout in the United Kingdom was constructed in 1909 and officially opened in 1910 as "Sollershott Circus" (Fig. 1.9).

Fig. 1.8 Place l`Étoile, Paris—old postcard

Fig. 1.9 The first roundabout in the United Kingdom—"Sollershott Circus", Letchworth, c. 1910 [1]

When looking at the earliest concepts of roundabouts, one is struck by features that are very similar to those fairly recently introduced as novel at modern roundabouts. Firstly, they are circular with significant deflection and curved deflection islands. Secondly there is the suggestion of dedicated right turn paths in Henard's design and thirdly, an over-run strip (truck apron) around the central island at Letchworth [1].

References

1. Brown, M. (1995). *The Design of Roundabouts*. UK: TRL, HMSO.
2. Pogačnik, A. (1980). Urbanistično planiranje, Univerza Edvarda Kardelja v Ljubljani, Fakulteta za arhitekturo, gradbeništvo in geodezijo, Ljubljana.
3. Marinović-Uzelac, A. (2001). Prostorno planiranje, Dom i svijet, Zagreb.
4. Milić, B. (2002). Razvoj grada kroz stoljeća—3. Dio, Novo doba, Školska knjiga, Zagreb.

Chapter 2
First Developments of Different Roundabout Types

2.1 Introduction

In the years from 1913 to 1914, Hellier suggested circular traffic systems at places, where several main roads would meet and the main connection of the circular system would prevent overload. At the conference of the local governmental committee on the subject of main roads in 1914, this idea was accepted as positive under the condition that the traffic requirements were met (every intersection should have sufficient empty space, and lawns alongside the intersections would be desirable). The initial phase of development in Europe was interrupted by the First World War. When the British Road Transport Board was set up in 1918, it was suggested that the roads of France should be the model for Europe. Gyratory systems were also used in the USA but there was great difficulty in regulating traffic, local ordinances were unenforceable and flouted, and there was no uniform rule of the road throughout the country. In 1924, at a US national conference, rights of way at intersections, and warning and stop signs were proposed. The "circus" idea continued to spread in the United Kingdom and was frequently recommended for busy junctions of more than four roads. During 1925–1926 a lot of gyratory systems were introduced in London. These were simply one-way systems around existing squares with fairly sharp corners. Unfortunately some of the important principles implied in Henard's concept, e.g., the entries into gaps during circling, operating over a short distance, were being lost. The transfer of these movements to a straight road caused differences in speeds at the conflict points but this may at first have been unimportant when all traffic speeds were quite low. The design was based solely on commonsense and experience [1].

Notwithstanding the foregoing, we could say that the first serious study of roundabouts began after the First World War and lasted until the late thirties of the last century. At that time it created many new ideas, some of which were implemented within modern roundabouts.

T. Tollazzi, *Alternative Types of Roundabouts*, Springer Tracts on Transportation and Traffic 6, DOI 10.1007/978-3-319-09084-9_2

The second period of in-depth research began in the fifties and lasted until the late sixties of the last century. During this time many new types of roundabouts were created which are still being implemented. This period also included a change of the priority rule, which completely changed the designs of roundabouts.

2.2 Trend to Non-circular Islands

The trend of using roundabouts was formally recognized in 1929, when collaboration between the Ministry of Transport and the Town Planning Institute of the United Kingdom resulted in the issue of the first "guidelines", which recommended that at crossings of one or more major roads, space should be provided for traffic to circulate on the "roundabout" system, and thus gave general guidelines for roundabout design. According to some sources [1], this was the first official use of the term "roundabout".

However, no doubt influenced by the conversions of many square and rectangular spaces to roundabouts, there was an assumption that a flat-sided central island shape was essential for the weaving of traffic, which was observed to take place on the outer sides. Splitter islands were made narrower and polygonal central islands were to have sides with minimum lengths of 110 ft., matching the number of entries, in order to "allow the traffic to sort itself out" [1]. The width of the circular carriageway was set at up to 40 ft. and a 30 ft. radius was declared for the central island corners. These suggested layouts included a four-arm (Fig. 2.1), and a six-arm roundabout with a hexagonal central island (Fig. 2.2) [1].

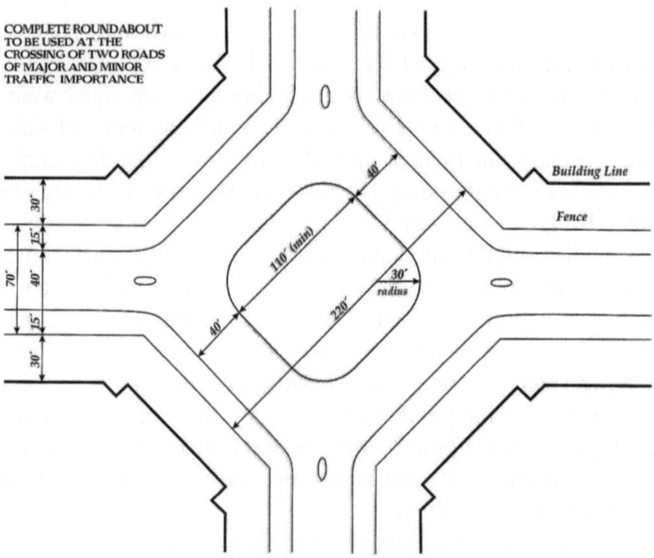

Fig. 2.1 Suggested layout of the four-arm roundabout [1]

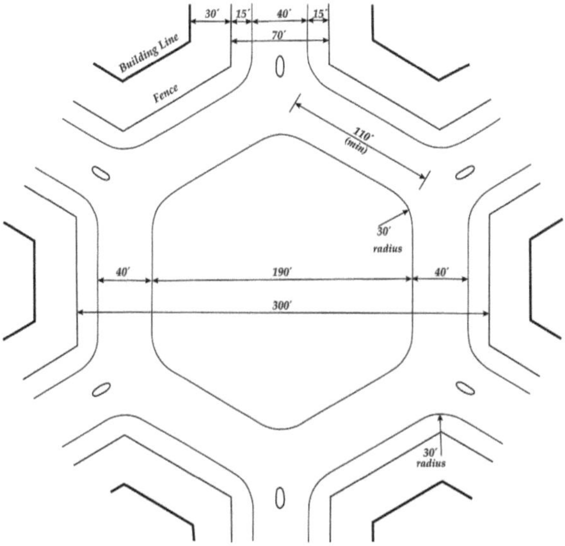

Fig. 2.2 Suggested layout of the six-arm roundabout [1]

Fig. 2.3 Piccadilly circus, original form—old postcard

This idea led changes to some of the existing roundabouts; circular central islands (Fig. 2.3) were replaced by hexagonal central islands (Fig. 2.4). Interestingly, it should be mentioned that the polygonal central islands of roundabouts were used again in the fifties of the last century, but only for a short period.

Fig. 2.4 Piccadilly circus, changed form—old postcard

By the mid-thirties, roundabouts were included in plans for solving traffic issues in the centers of many cities. In 1933, Watson set the following priorities: decreases of congestion, timely synchronized driving through the intersection, more comfortable traveling, safer traffic flow, reduction or complete elimination of police traffic controls at intersections and reductions of interferences in the courses of traffic [1]. According to Watson, the main downsides were inappropriateness and danger to pedestrians, and the danger of shoplifting in those cases where there were many heavy vehicles on the roundabout driving passed shops on the central island. With the increase in traffic, the number of traffic accidents also increased in the United Kingdom. In addition to increasingly faster motor vehicles, horse-drawn carriages were also part of the traffic as well as a large number of cyclists and pedestrians. At that time, there were no specific traffic rules applicable for pedestrians crossing a street. Later on, streets with two-directional carriageways were proposed, with dividing lanes in the middle, emergency lanes, cycle lanes and corridors for pedestrians, thereby preventing conflicts occurring due to the different speeds. These guidelines are indicated in the layouts of roundabouts in 1937 (Fig. 2.5) [1].

It is necessary to point out that some roundabouts from that time still exist, and carry out their roles very well (Figs. 2.6 and 2.7).

Later on, there was a proposal to use a round central island, planted with bushes, thus preventing pedestrians crossing the central island. This idea generated a new form of roundabout. Roundabouts acquired slightly expanded entries

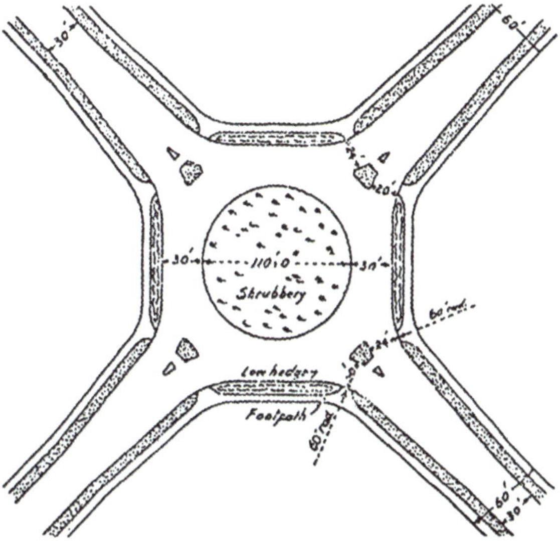

Fig. 2.5 Typical layout of roundabout according to MoT circular 390 [1]

Fig. 2.6 Square roundabout with square island; Coventry

Fig. 2.7 Square roundabout with circular island; Coventry

for easier turning to the left (in the UK), curved entries and splitter islands at the entries, containing marked pedestrian crossings and exits larger than the entries (faster–easier exit from the roundabout) [1].

2.3 The Period of Intensive Experimentation with New Layouts

As said before, the second period of in-depth research began in the fifties and ran until the late sixties of the last century. During this time many new types of roundabouts were created which are still being implemented.

By 1966 the situation at peak hour congestion and control at roundabouts had become intolerable. It needs to be stressed that the first roundabouts differed from recent roundabouts in the right-of-way rule. The vehicle at the entry had the right-of-way over the vehicles in the circular flow. This resulted in large radii of roundabouts and with narrow splitter islands—all with the purpose of acquiring the longest possible circular segments, where the weaving maneuvers of traffic flow took place. The dimensioning of roundabouts of that time and the calculation of capacity were based on Wardrop's definition of practical capacity [1], coming from the capacity of a circular segment between consecutive entries, where the weaving of vehicles' maneuvers took place. The deficiencies of this kind of traffic management at roundabouts began to show with the growth of motorized traffic. By giving priority to the vehicles at the entries, the vehicles on the circular carriageway began to pile up. Due to the increased motorized traffic, the traffic at these intersections came to a complete halt, since any potential queue at one of the circular segments would block the operation of the entire roundabout. Therefore, due to rapidly increasing traffic in the seventies in the UK, the blockage phenomenon at standard roundabouts caused a lot of confusion within the traffic system, in particular at places where uneven flows at fast entries caused the slower circulating vehicles to give priority to the faster entering vehicles. The uninterrupted entry-flow caused interruptions in the circling of traffic in the circular flow, and thus congestion.

New forms of layouts were created for roundabouts at the more overloaded access roads in London. One of the proposed solutions was also a change in the traffic regime—the right-of way. For this purpose, they conducted a few experiments under real conditions at the existing roundabouts. General application of this rule of taking away the priority of the vehicles at the entry became effective in November 1966, and over a few months actually changed the concept of roundabouts totally. It eliminated the problem of congestion, improved capacity, diminished the number of traffic accidents and at the same time caused a complete change in the philosophy of operating performance and the designing of roundabouts. By giving the right-of-way to vehicles in the circular flow, the problem of a roundabout's capacity was transferred from the weaving area to the area of entries onto a roundabout. This caused the need for widening access roads at the entries, while the size of the central island began to lose its meaning regarding practical capacity [1].

The consequences of the new rule-of-way were diminished roundabouts of the same capacity (less required space), increased traffic safety, and roundabouts' blockages at much higher traffic loads. Some layouts of the roundabouts of that time are presented in following.

2.3.1 One-Lane Roundabout

The standard one-lane roundabout (Fig. 2.8), which is the more numerous type of roundabout all over the world, has only one lane at each of the entries and exits, as well as on the circulatory carriageway (or roadway). For pedestrians and motorized vehicles, this type of roundabout seems to be the safest type amongst all types of "classic" or "standard" grade intersections.

The dimensions of the outer diameter differ from country to country, but is usually between 26 m (as a minimum; better 29 m) and (in some countries) 45 m.

A standard one-lane roundabout has a central island, made up of two parts: the traversable (truck apron) and non-traversable parts. The center of a modern one-lane roundabout provides a visual barrier across the intersection for the drivers entering it (Fig. 2.9). These functions assisted the drivers when focusing only on the traffic coming towards them along the path of the circle (and non-motorized participants).

Fig. 2.8 Typical Slovenian one-lane roundabout

Fig. 2.9 Visual barrier across central island; sculpture of wales; south France

Globally, pedestrians are prohibited (except in Mexico (Fig. 2.10), Vietnam and a few other countries) from entering the central islands of roundabouts, but there exist also some differences in the case of assembly roundabouts (Fig. 2.11).

Fig. 2.10 Roundabout with pedestrian crossing into central island; Teotihuacan, Mexico

Fig. 2.11 Assembly roundabout with motorbikes' parking at central island; near Italian border with France

Due to the needs of larger vehicles (swept path for turning) the circular carriageway must be wider than the usual lane. Having only 26 m diameter, the circular lane must be wider by up to 8 m (and at 29 m diameter 6.5 m is enough—if it includes a traversable ring—truck apron). This type of roundabout can be applied in urban as well as in rural areas.

Deflection on entry is used to maintain low speed operations at roundabouts. Drivers must maneuver (are "deflected") around the central island, at speeds of 25–40 km/h.

For pedestrians the walkway crossings (usually zebra crossings, which impose an absolute right of way for pedestrians) the entries and exits should be built at distances of 5–10 m from the margin of the circle (because of the waiting spaces at the entrances and exits).

While roundabouts can reduce accidents overall compared to other junction types, crashes involving cyclists may not experience similar reductions for some designs. Looking globally, the only remaining significant risk at a single-lane roundabout (in some countries) involves cyclists. Accordingly, the cyclists' lanes in the guidelines for roundabouts' designs differ from country to country, but here we can point out three different types of cyclists' management (Fig. 2.12).

In continental European countries (except Holland), painted cycle lanes at the peripheral margins of circulatory carriageways are not allowed since they have proven to be very dangerous for cyclists. In Germany (and some other countries e.g., Slovenia) with a traffic volume of up to about 15,000 veh/day, cyclists can be safely accommodated on the circular lane without any additional installations in urban areas. In Germany, even if the approaching lanes of the roundabout are equipped with separated cycle tracks, the two-wheelers are guided to the normal traffic lane at the approach in order to guide the cyclists through the single-lane roundabout and lead them back on to a cycle track after leaving the intersection, towards the desired direction [2].

Above a volume of 15,000 veh/day, separate cycle paths (three types) are regarded as being useful in most countries. These, however, should also have a distance of around 5–10 m from the circle at the point where they are crossing the entries and exits. Shorter distances have negative impacts from the traffic safety and capacity point of view. At closer distances the visibility regarding cyclists is impeded for the drivers of trucks, and waiting places (between the margin of the

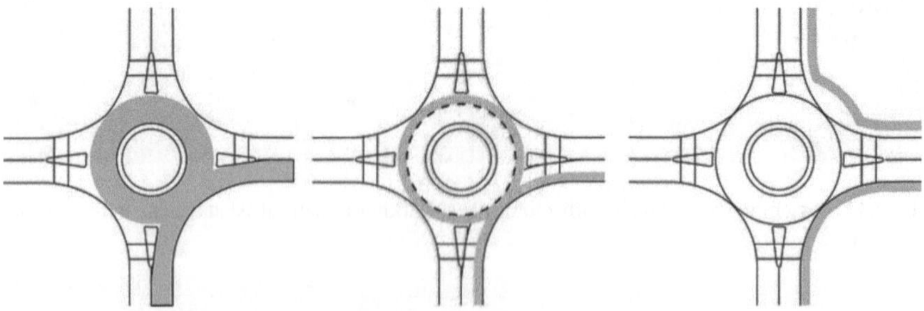

Fig. 2.12 Three different types of cyclists' managing

circle and the inner edge of a zebra crossing) at entries and exits have a strong influence on the capacity.

If the adjacent footpaths of a roundabout are improperly designed, there is increased risk for persons with visual impairments. This is because it is more difficult (than at a signalized intersection) to detect by hearing whether there is a gap in the traffic adequate enough for crossing.

Lighting is very important, even though there are countries in which the lighting of roundabouts is optional. There are two types of lighting on roundabouts; on the central island and out at the peripheral margin of the circulatory roadway. Some countries have had bad experiences with lighting pools at central islands (because incoming drivers were being blinded by the light).

The figures about the capacities of one-lane roundabouts differ from country to country (human behavior), but may be expected to handle approximately 20,000–26,000 veh/day. It is very important to know that under several traffic conditions, a roundabout may operate with less delay than an intersection with traffic signal control or all-way stop control. Unlike an all-way stop intersection, a roundabout does not require a complete stop by all entering vehicles, thus reducing both individual delay and delays resulting from vehicle queues. A roundabout may also operate much more efficiently than a signalized intersection because the drivers are able to proceed when traffic is clear without the delays that occur while waiting for traffic signals to change. It also needs to be stressed that these advantages also reduce air pollution from many idling vehicles waiting for traffic lights to change, which is a very important criterion in residential areas.

2.3.2 Square Roundabout

We can distinguish between two types of square roundabouts according to their origins. The first type originated from those initial old town squares with four or more intersecting roads, initially intended only for horse-drawn vehicles and pedestrians (Fig. 2.13).

Later it became necessary to separate pedestrians from motorized traffic. Thus, at the outer edges of the square, a circulatory carriageway was created, intended only for motorized vehicles, while the remaining part became an elevated platform intended for pedestrians only. These square roundabouts at their inceptions did not include splitter islands neither pedestrian crossings (Fig. 2.14).

As a rule, they were round or oval, often containing trees, grass, sculptures, fountains, and benches for recreational use.

At the beginning, there were no traffic rules, and square roundabouts were without splitter islands and pedestrian crossings. Later on, there was a need to introduce traffic rules. First, the vehicles at the approach had the right of way and splitter islands and pedestrian crossings were initiated.

It is necessary to stress that many square roundabouts from that time still exist, and that they carry out their roles very well. Nowadays, they are usually signalized

Fig. 2.13 Town square; Maribor, Slovenia—old postcard; about 1890

Fig. 2.14 Town square—60 years later; Maribor, Slovenia—old postcard; about 1950

because they are mainly located in city centers (large traffic volume) or because they are intersected by the subsequently implemented suburban railways. These types of square roundabouts can be found all over Europe in the older and larger towns (Fig. 2.15), and also in the rest of the world.

The second group of square roundabouts is more recent. Usually they occurred because the traffic regimes on the existing roads had changed. The basic

Fig. 2.15 Prague, Czech Republic

characteristic of this group of square roundabouts is that they do not have a round central island, but a rectangular or a square one. In general, there are two types: those with tangential approaches and those with combinations of radial and tangential approaches.

A square roundabout with tangential approaches (Fig. 2.16) deals with two-lane approaches only (and are intended for two-way traffic), while the circular roadway can be two- or three-lane (and are intended for one-way traffic only). Sections between one approach and the following exit must be sufficiently long, so that weaving maneuvers can be performed. The straight parts of a square roundabout enable high speeds; therefore, we must devote a special attention to the traffic management of the non-motorized participants. As a rule, the management of non-motorized participants is implemented on another level or by traffic lights. When considering tangential access, the traffic regime with a yield traffic sign is not appropriate, therefore most of these square roundabouts are equipped with underpasses at least if not with traffic lights together with appropriate public lighting.

At a square roundabout with a combination of radial and tangential approaches (Fig. 2.17) we are dealing with several two-lane approaches (intended for two way traffic) and several one- way approaches (intended for one-way traffic only). As a rule, the circular roadway is one-lane; however it can also be two-lane (intended for one-way traffic only). In the cases of one-lane circular roadways, there is no need for long straight parts because there is no weaving of traffic flows. There are also cases of two-lane roundabouts where the right-hand lane is intended for parallel parking.

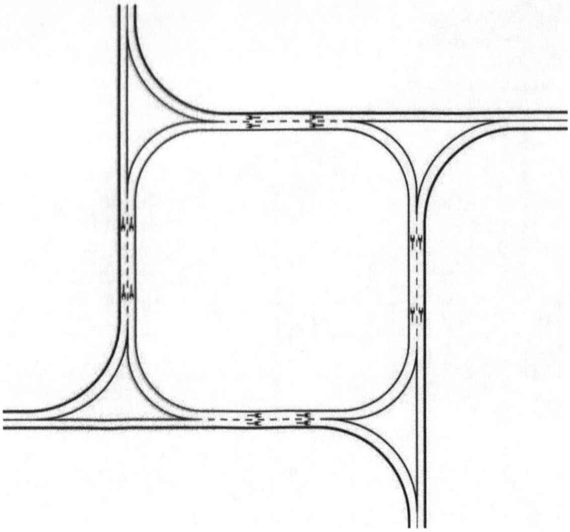

Fig. 2.16 Square roundabout with tangential approaches

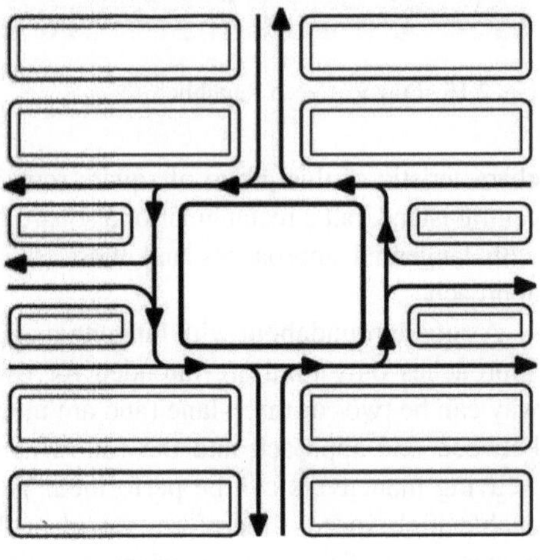

Fig. 2.17 Square roundabout with a combination of radial and tangential approaches

This type of square roundabout includes exclusively entering, exclusively exiting or combined (entry and exit) approaches. The numbers of each individual type of approach depend on traffic needs.

This type of square roundabout does not present as much danger for non-motorized participants as square roundabouts with tangential approaches, since fewer approaches are tangential (while others are radial). Combined approaches are radial (T intersections) and slow down the traffic, which provides a higher level of traffic safety. A T-intersection (90° turning left or right) is a natural traffic calming measure, and because of that cyclists and pedestrians are not in conflict with motorized vehicles at these points.

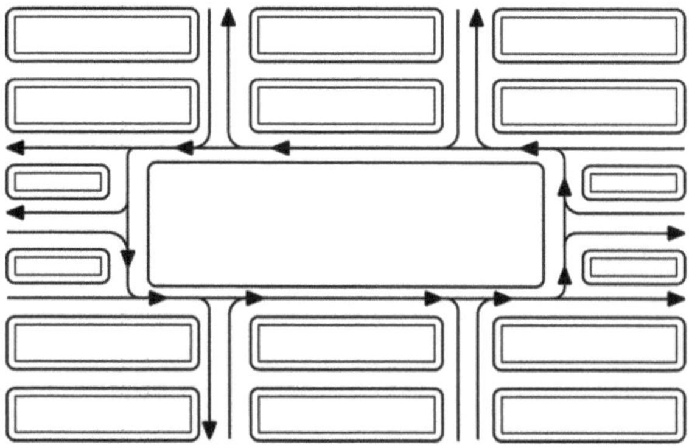

Fig. 2.18 Square roundabout with a combination of radial and tangential approaches on two squares of streets

However, we must pay attention to the navigation of pedestrians and cyclists through one-lane approaches. The speeds may be higher because of tangential approaches; therefore traffic calming measures can also be implemented as well as traffic signalization of the whole square roundabout, even though under well-regulated conditions it is also possible to implement a yield traffic regime.

It needs to be emphasized that this is a very good solution in everyday life, when the systems of the streets have already been built, when it intersects perpendicularly (radial intersection of several streets is rare), and when it can no longer be changed because of the build-up surrounding it. We must also emphasize that a square roundabout of this shape can be implemented not only with one square of streets but also with two or more (Fig. 2.18).

The capacity of a square roundabout with a combination of radial and tangential approaches is lower compared to the first type; however, the level of traffic safety is higher (lower speeds, no weaving, half of the approaches are radial…).

The inside (central islands) of both types of square roundabouts are easier to access compared to the standard roundabout, regardless of whether we are dealing with a signalized pedestrian crossing or an underpass (according to the author's experience, pedestrian crossings leading towards central islands have been implemented in a few countries only).

2.3.3 Large Roundabout

Roundabouts were previously designed on the assumption that the weaving movements of vehicles took place on the circulatory carriageways. The early assumptions of roundabouts' designs were that larger roundabouts would carry more traffic than smaller ones of similar shapes and that the longer sides of the "weaving sections" would improve capacity. Roundabouts with an inscribed diameter well in excess of 100 m were suggested in 1945 for typical arterial road peak traffic flows. Before the change in the priority rule (and even for some time afterwards) large roundabouts were preferred for providing protection against "lock-up" at traffic peaks [1].

Fig. 2.19 Large roundabout; from HCM 1950 [2]

In the US, in the years following the World War II, roundabouts in urban areas were hardly used, and outside urban areas roundabouts were merely an exception and not a rule. The advice during planning was that, due to the deficiencies of roundabouts or rotaries, their application should be limited. They were more often used as a solution to the crossings of a greater number of roads in rural areas with enough space for their application (Fig. 2.19). Roundabouts gradually replaced the standard intersections, while their advantages were nullified by the increasing traffic in the US. At that time the idea of multiple level intersections came to life.

Roundabouts of that time were of large dimensions, especially on high-speed roads. It was expected that roundabouts should not significantly influence the speed of vehicles. The result was roundabouts of stretched (oval) shapes, which gave priority to the transit traffic and emphasized the right-of-way of the traffic at the entry. Several such intersections are also still in use in European countries (Fig. 2.20).

The formation of "weaving sections" at the roundabout was taken from the weaving of traffic between the connecting directions of a cloverleaf (common ramp for entry and exit). This method of calculation showed that the weaving maneuver was as a principle, dependent on the lengths and widths of the weaving areas and on the structures of the traffic. Planned speeds for the weaving areas at the roundabouts of that time were from 65 to 110 km/h. The belief at that time was that roundabouts with low speeds would amount to a low-level of traffic service,

Fig. 2.20 Large roundabout of stretched—oval shape, León, Spain; postcard

because it would be impossible to reach a sufficient number of weavings. Such an opinion still prevailed in 1965, when creating the Highway Capacity Manual [3].

Presently, large roundabouts are no longer constructed, as they are outdated for several reasons. Large roundabouts, especially those with faster traffic, are unpopular with some cyclists. This problem is sometimes addressed at larger roundabouts by taking foot and cycle traffic through a series of underpasses or alternative routes. In rural areas, large roundabouts require huge space and long splitter islands further increase the cost. The same situation occurs in urban areas—large roundabouts "eat up" a lot of urban space. Temporary widening and outside diameter space requirements increase the running costs of construction as well.

Therefore, new large roundabouts are no longer constructed presently, while the existing ones are being reconstructed into some more suitable types of intersection in terms of traffic safety. This means smaller entrance radii, reductions in the number of lanes at entries and exits and along the circulatory roadway, or the provision of traffic signals at such roundabouts (Fig. 2.21).

2.3.4 Double-lane and Multi-lane Roundabouts

The basic reason for the construction of two- and multi-lane roundabouts was an expectation that the capacities of such roundabouts would be duplicated.

Fig. 2.21 Large roundabout with traffic signals; Berlin, Germany—postcard

Consequently, roundabouts of two shapes were implemented in the past: with two lanes at entries/exits and as a single lane along the circulatory carriageway, or with two lanes at entries/exits and two lanes along the circulatory carriageway; the purposes of which were to increase (duplicate) their capacities. During the early seventies of the past century, there were a large number of such roundabouts constructed in the UK, and also in other European countries later on (Fig. 2.22).

After a while, the reality contradicted all the technical assumptions and numerical calculations that supported the introduction of such a type of roundabout.

Let us start at the beginning. The main advantage of a one-lane roundabout, compared to the "standard" intersection, is the elimination of conflict spots (somewhere also "points") of the first (crossing) and second (weaving) grade, and the reduction of conflict spots of the third grade (merging, diverging). Theoretically, the classic four-arm (somewhere also "four-leg") intersection has 32 conflict spots (16 crossing, 8 merging and 8 diverging), while the one-lane four-arm roundabout has only 8 conflict spots (4 merging and 4 diverging) (Fig. 2.23).

If there are two circular lanes, the number of conflict spots increases by the number of weaving conflict spots, which theoretically equals the number of arms, however, this number is still lower than 32 (Fig. 2.24). But, from the practical point of view, we are not only speaking of conflict spots at the multi-lane roundabouts, but also of conflict sections (sequences of conflict spots), since there are no predetermined spots for drivers where they must change lanes along the circulatory carriageway.

Fig. 2.22 Multi-lane roundabout; Ljubljana, Slovenia

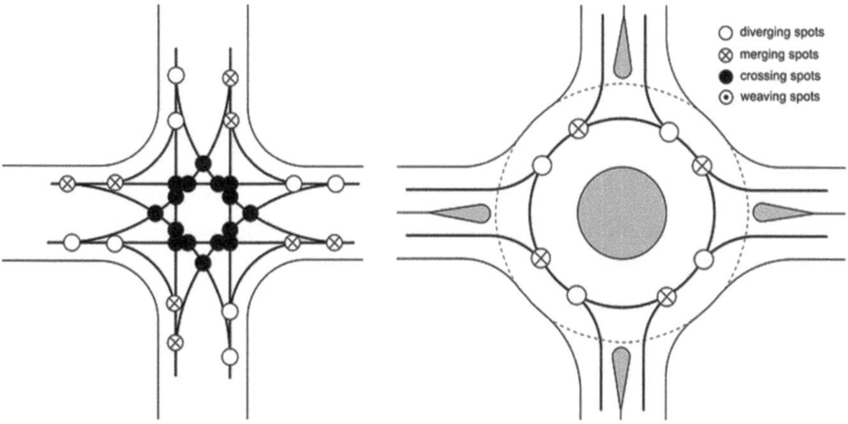

Fig. 2.23 Conflict spots at a four-arm "standard" intersection and a four-arm roundabout

At multilane roundabouts with two-lane entries and exits, the traffic-safety conditions are even slightly worse (Fig. 2.25). In this case, there are conflicts at the spots of crossing the circulating lanes at the entries and even bigger in the course of changing traffic lanes along the circulatory carriageways. However, by far the most dangerous is the maneuvering when leaving the roundabout.

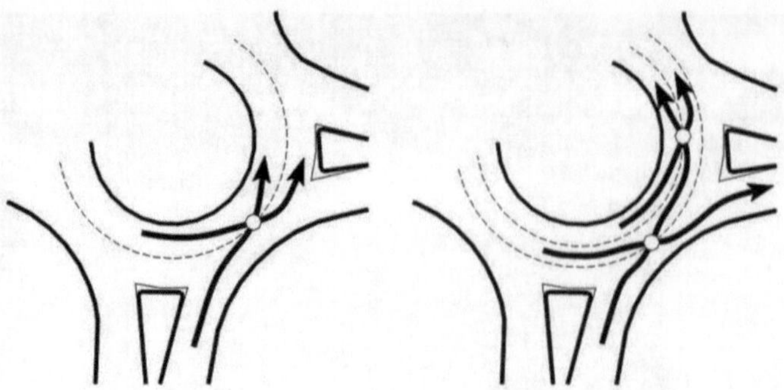

Fig. 2.24 Conflict spots at the multi-lane roundabouts with one entry lane

Fig. 2.25 Conflict spots at multi-lane roundabouts with two-lane entries and exits

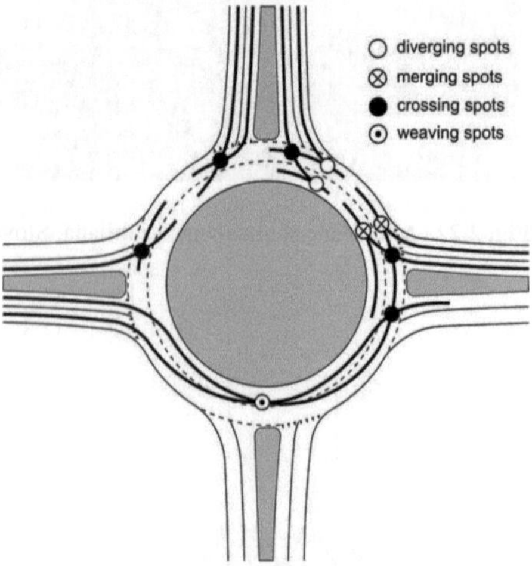

Conflict spots at multi-lane roundabouts with two-lane entries and exits are located at the following areas:

- roundabout approaches (weaving, when approaching the roundabout);
- entry onto the roundabout (and crossing a circulatory traffic flow);
- multilane circulatory carriageway (weaving in the course of changing traffic lanes);
- leaving the roundabout (and crossing a circulatory traffic flow);
- roundabout approaches (weaving, when driving away from a roundabout).

It is necessary to stress that it is possible to reduce the numbers of some conflict spots with certain measures; however some types of conflict spots cannot be eliminated because they are characteristics of the roundabouts' types.

As mentioned before, the main reason for the introduction of these types of roundabouts was based on expectations that their capacities would increase substantially. Later on, real-life contradicted all the technical assumptions and numerical calculations. Namely, in many countries, it was figured out that the second lane along the circulatory carriageway contributed to increased capacity only by an additional 30 % (and not by 100 % as originally expected and mathematically predicted in a rather illusionary way). The fact that the second lane in the circulatory roadway increased the capacity by only 30–40 % was replicated in several countries; however, we would only point out Austria, Lithuania [4], Germany [5] and Slovenia [6]. It was also discovered that on these types of roundabouts, the level of traffic safety was significantly lower than at one-lane roundabouts. There were many reasons for that. One of the more important reasons was surely the fact that in the past, two-lane roundabouts that were too small were being constructed in those countries which contradicted the statutory rule of the mandatory use of inner circulatory traffic lanes in those cases when the driver does not leave the roundabout at the next exit (an average driver did not have sufficient length to change the driving lane along the circulatory carriageway). The second reason was that the inner circulatory traffic lane along the circulatory carriageway was avoided by younger and more senior drivers as they felt insecure when changing.

Over the years, these types of roundabouts have gained a bad reputation regarding their safety and their limitations regarding capacity which (in spite of the large consumption of space) did not exceed an ADT beyond 40,000 veh/day.

Following the above-mentioned findings and the fact that we are now familiar with newer and safer types of two-lane roundabouts (with significantly larger capacities and levels of traffic safety), some countries (e.g., The Netherlands, Slovenia) have even forbidden the constructions of new "standard" two-lane roundabouts in their recent regulations. The existing two-lane roundabouts in these countries are being reconstructed into some safer two-lane roundabout types (e.g., into turbo-roundabouts). As it seems the only exception is the compact semi two-lane roundabout in Germany (Fig. 2.26). As a first approach towards larger roundabouts, a new intersection type has been created, the compact semi-two-lane circle. The ideas for this type were first described in a German preliminary

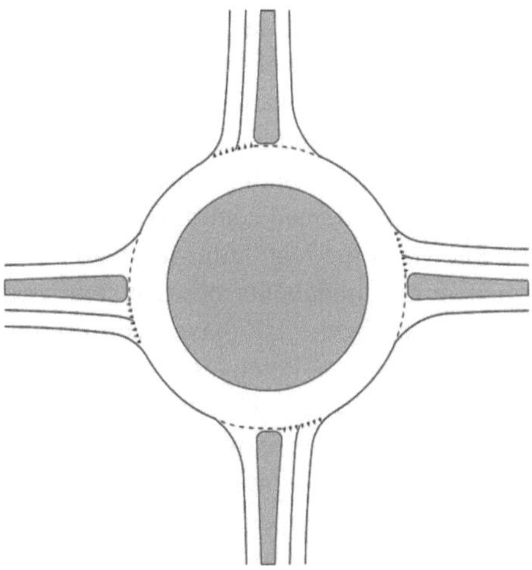

Fig. 2.26 Compact semi-two-lane roundabout; sketch

document in 2001, accepted as guidelines from 2006, and consequently these new types were built by several municipalities and highway authorities [5]. The design of compact two-lane roundabouts is similar to the concept of one-lane roundabouts. The main difference is the width of the circular lane. It is wide enough for passenger cars to drive along side by side, if required. However, the circle lane has no lane markings. Large trucks and buses are forced to use the whole width of the circulatory carriageway when making their way through the roundabout.

The outer diameter of this type of roundabout is 40–60 m, the circle lane width being 8–10 m, and without lane markings (to prevent drivers from overtaking). The number of entry lanes is in accordance with traffic volumes (one- or two-lane), but always has only one-lane exits. It is necessary to point out that no cyclists are allowed on the circulatory carriageway. These roundabouts are today state-of-the-art solutions in Germany.

2.3.5 Mini-Roundabout

Information on where and when the first mini-roundabouts in the world were implemented is inconsistent. However, the widespread belief is that the "traffic circles", as applied by Eno in the first decade of the 20th century in America were the first real mini-roundabouts [1].

The traffic used to circle around a central pole or a stone tower known as the "dummy cop", and the traffic ran in one direction. Later on, small mushroom-shaped islands started to be introduced, mostly on roundabouts with a minimal diameter of 10 m. Some old documents show that the state of Connecticut (where Eno was from) was the first and maybe the only administrative area that used a central island, marked merely by white coloring. Eno's mini-roundabout with a "target" in the center was composed merely of a central circle, 60 cm wide and marked by white coloring, surrounded by two white concentric circles, 30 cm wide, situated at a distance of 30 cm. The total diameter thus amounted to 3 m [1]. This was the origin of the target type of mini roundabout still widely used in the UK (Fig. 2.27), Malta and Australia.

Later, studies and analyses on the suitability of mini-roundabouts' applications were performed, which resulted in guidelines (in 1971) on mini-roundabouts with flat or slightly domed central islands having radii of up to 4 m. These roundabouts are known as the real "mini-roundabouts". The experiments continued until 1975, when mini-roundabouts became regulated by law in the UK. Since then, mini-roundabouts in the UK have been used in urban areas but only exceptionally in transitory areas. Thereby the operation of intersections situated within a small space was improved. The first roundabouts were implemented only for three-arm intersections and were only later introduced at intersections with four arms.

The main use of mini roundabouts in the UK is during the conversion from other intersection types, including traffic signals. The main criterion for safe operation is an appropriate speed of vehicles at all entries into a roundabout, which

Fig. 2.27 The origin of the mini roundabout, still widely used; Salisbury, the UK

should amount to a maximum of 50 km/h. If the central island has a diameter of 4 m or less, no raised island (or street furniture) is permitted on it, in order to allow long vehicles to over-run. In these cases, the central island is slightly domed and painted with white reflective paint.

Mini-roundabouts in the UK have been presented by many authors, particularly by those UK experts who have been intensively involved in the development of the UK mini-roundabouts for decades, and any new findings are immediately transferred to new standards and guidelines. It is especially worth mentioning Clive Sawers who is—in the perspective of being a "non-Englishman"—probably the best expert in the area of mini roundabouts in the UK and his books have been read all over the world. "Mini-roundabouts—A Definitive Guide" [7] is essential reading for all engineers, designers and traffic safety auditors practicing in this field and important too for planners and town centers designers. The book contains sound advice on site selection, layout details, and crossroads, a simple capacity test and much guidance on features of design that contribute to traffic safety.

The last standards for the geometric designing of mini-roundabouts were published in 2007 [8]. A typical UK mini-roundabout has to have a central island, composed of a circular solid white road marking between 1 and 4 m in diameter that is capable of being driven over (see Fig. 2.28). A vehicle proceeding through the mini-roundabout must keep to the left of the white circle unless the size of the vehicle or the layout of the mini-roundabout makes this impractical. Although the standard is nominally intended for trunk roads, there are very few mini-roundabouts on such roads. It is very important to stress that UK mini-roundabouts should only be used

Fig. 2.28 Recently designed mini-roundabout; Hendon, London

on urban single carriageway roads where the speed limit is 30 mph or less, and the 85th percentage dry weather speed of traffic is less than 35 mph within a distance of 70 m from the give way line. They are seen as a remedial measure for a poorly performing priority junction rather than a junction type in their own right [9].

The last UK standards for the geometric design of mini-roundabouts states that mini-roundabouts must not be used at new junctions or where the traffic flow on any arm is less than 500 veh/day. Four-arm mini-roundabouts should not be used if the total inflow in the peak period exceeds 500 veh/h. No mini-roundabouts should have five or more arms, although double mini-roundabouts may be used at a pair of closely spaced priority junctions.

In general, UK mini-roundabouts are not considered as being speed reduction measures as such, but are suitable for use as part of an urban traffic calming scheme. Because mini-roundabouts were previously designed according to the roundabout standard, they follow the same general principles, often having entries which flare into two (narrow) lanes (which is unique regarding other European countries when creating mini-roundabouts), and because of this the inscribed circle diameter should not exceed 28 m.

Splitter islands may be curbed or may be created using road markings (just painted). They must be curbed (Fig. 2.29) where otherwise vehicles would find it easier to pass on the wrong side of the white circle. Deflection by the white circle is not essential, but a lateral shift at the entry of at least 0.8 m, normally on the offside, is considered good practice.

Fig. 2.29 Curbed splitter island; Cambridge

The circles of UK mini—roundabouts may be domed to deter light vehicles and to improve conspicuity (in most other European countries creating mini-roundabouts, the term "may" is replaced by the term "must"). The maximum recommended height at their centers is 10 cm for a circle of diameter 4 m, with smaller diameter domed circles reduced pro-rata. What is very interesting is that domed circles should not be used if they are likely to be run over by buses, thus avoiding possible discomfort to passengers. What is also need to be stressed about traffic safety is that models for safety at their mini-roundabouts were developed a long time before [10]. Some links report that today there are about 5,000 mini-roundabouts around the UK and a great deal of experience has been gained from their application.

Today, this type of roundabout is also in frequent use in many other European countries. It has proven to be a very good experience in e.g., Germany, France, Austria, Switzerland, Slovenia, and Croatia. These countries, however, have slightly changed the original mini-roundabout layout by altering conditions for their implementations, provided different traffic signs and the layouts of the central islands. Therefore, we present below the experiences in some other European countries. First, it is necessary to point out the fact that at present each country creating mini-roundabouts has adopted its own guidelines for mini-roundabouts' design, where such rules might be substantially different.

Germany has now 25-years of experience with different types of modern roundabouts, including mini-roundabouts. Experiments with mini-roundabouts in Germany began in 1995 in the state of Northern-Westphalia, and the leader of that experiment was Prof. Brilon [11]. Experiments included 13 intersections that were

Fig. 2.30 Typical German mini-roundabout; suburb of Bochum

converted from non-signalized intersections into mini-roundabouts (Fig. 2.30). The success was overwhelming. They could carry up to 20.000 veh/day without major delays to vehicles, they could easily be built—sometimes without significant investment costs—and they turned out to be very safe [5].

German mini-roundabouts can be applied only in urban areas and have inscribed circle diameters of between 13 and 24 m (measured between the curbs), the circulatory carriageway widths are between 4.5 and 6 m, and with cross-slopes of 2.5 %, which must be inclined towards the outside. It is not sufficient to establish central islands just with some road markings, so central islands have—in the center—maximum heights of 12 cm above the circular line. In order to convince drivers to accept the roundabout driving rules, a minimum curb height of 4 or 5 cm has been identified from experience.

Experiments with rural mini-roundabouts have also been performed. As a result, mini-roundabouts are no longer recommended outside built-up areas, due to safety reasons.

As an interesting feature, it is worth mentioning that they provide their mini-roundabouts with a special traffic sign indicating a mini-roundabout (Fig. 2.31), which is, according to the author of this book, a very good idea.

We should also mention that new research on mini-roundabouts is continuously going on in Germany [12].

Good experience with mini-roundabouts is also observed in France, both three- (Fig. 2.32) and four-arm mini-roundabouts. Mini-roundabouts in France are in

Fig. 2.31 Traffic sign in mini-roundabout; suburb of Cologne

Fig. 2.32 Typical French three-arm roundabout; Provence

Fig. 2.33 Domed circle and curbed splitter islands; suburb of Nice

frequent use, created usually in urban areas at locations with limited space options. Splitter islands may be curbed or may be created using road markings (if there is enough space they are curbed).

The circles of the French mini-roundabouts need to be domed in order to deter light vehicles and to improve conspicuity (Fig. 2.33), and deflection is also very important at French mini-roundabouts.

Almost the same situation exists in Slovenia (Fig. 2.34). Their mini-roundabouts (both three- and four-arm mini-roundabouts) can be applied only in urban areas and have inscribed circle diameters of between 13 and 24 m, the circulatory carriageway widths are between 4.5 and 6 m, and with cross-slopes of 2.5 %, which must be inclined towards the outside. It is not sufficient to establish central islands just with some road markings. Central islands of the Slovenian mini-roundabouts need to be domed, and—in the center—maximum heights of 12 cm above the circular line. In order to convince drivers to accept the roundabout driving rules, a minimum curb height of 3 cm has been identified from experience.

Good experience with mini-roundabouts is also observed in Italy, even at this moment they do not have their own guidelines for mini-roundabouts (Fig. 2.35). Mini-roundabouts in Italy are in frequent use, created usually in urban areas at locations with limited space options. Splitter islands are usually curbed, and circular islands are usually domed.

A little bit different situation is in The Netherlands. It seems they do not prefer mini-roundabouts, even though some have been built (Fig. 2.36). They have

Fig. 2.34 Typical Slovenian three-arm mini-roundabout; Maribor

Fig. 2.35 Italian mini-roundabout; Sanremo

Fig. 2.36 Mini-roundabout with painted central island; Maastricht, The Netherlands

none of their own mini-roundabout's guidelines, and their philosophy is that "the English term mini-roundabout does not refer to the outside diameter of the roundabout but to the diameter and the shape of the central island". They do not prefer mini-roundabouts for two main reasons:

- white painted central island does not function as drivers over-run it or find it easier to pass on the wrong side of the white painted central island;
- requirements of the deflection criteria are not met.

But, what they really prefer instead of a mini-roundabout is a specific type of neighborhood traffic circle, the so-called "punaise" ("pushpin" or "road stud type roundabout"), presented in Sect. 4.4.

2.3.6 *Double Mini-Roundabout with Short Central Link Road*

The test-track experiments commenced in 1967 were working on the basic principle of making better use of spaces available at junctions. Various outline shapes and methods of control were compared for a particular area of intersection widening. Although not entirely unexpected, the success of the experimental off-side priority control roundabouts featuring small islands and widely flared entries was very encouraging. This period of intensive public road experimentation with new layouts continued for several years in the UK, at least until 1972. The boundaries of driver acceptance were established in principle during this period. A range of applications were produced to suit various conditions. Initially mini islands were used at large roundabouts but not all of these had adequate deflection. Mini islands were found to be more successful at the urban intersections of small areas, as an alternative to priority junctions. Many other specialized layouts were developed at that time. Double and multiple island roundabouts and ring junctions were found to have advantages at some sites [1]. One of them is the double roundabout with a short central link road (with joint splitter islands), also called "closely spaced roundabouts", which is still in frequent use in some countries. This solution may be constructed as a double mini-roundabout (Fig. 2.37) or, alternately, as a two standard one-lane roundabout. Both solutions require standard dimensions for these two types of roundabouts. Accordingly, these roundabouts' dimensions are identical to the dimensions of individual roundabout types. Such a type of roundabout is most frequently located at an existing H intersection (i.e., two T or + intersections of a short distance).

Numerous roundabouts of this type (particularly with mini-roundabouts) from the early seventies are still used in the UK (Fig. 2.38).

It is interesting to mention that such a solution is used rather frequently in Croatia, especially recently (Figs. 2.39 and 2.40). Over recent years, namely, there have been quite a lot of such examples applied in urban areas of Croatia. In most cases the solution includes mini-roundabouts.

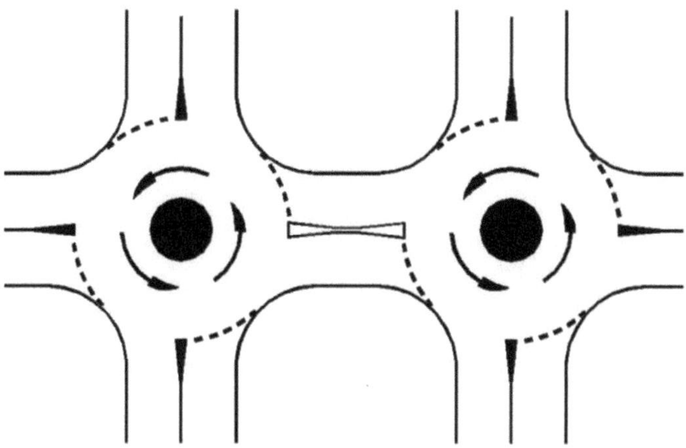

Fig. 2.37 Double mini-roundabout with short central link road; sketch

Fig. 2.38 Double mini-roundabout with short central link road; suburb of Coventry, the UK

Fig. 2.39 Double mini-roundabout with short central link road; city of Zagreb, Croatia

Fig. 2.40 Three arm double mini-roundabout with short central link road; island Rab, Croatia

2.3.7 Dumb-Bell Roundabout

During several past decades, ramp intersections were configured as "standard diamond interchanges", but some 30 years ago, the promotion of a new solution started, often called a dumb-bell roundabout (due to its aerial resemblance to a dumb-bell, a piece of equipment used in weight training) (Fig. 2.41).

The dumb-bell is a "hybrid" combining the diamond and the roundabout, which makes it a very close relative of both, as one is a direct descendant of the other. In short, it combines the capacity benefits of a (usually) one-lane roundabout with the smaller footprint and single bridge of a standard diamond junction.

A dumb-bell roundabout is a better solution than a "standard diamond interchange" because of several reasons. It can generally handle traffic with fewer approach lanes than other intersection types. This type of roundabout reduces construction costs by eliminating the need for a wider flyover [diamond—minimum three (usually four lanes), dumb-bell—just two lanes], and less space. As a rule, drivers within a "standard diamond interchange" driving at high speeds may

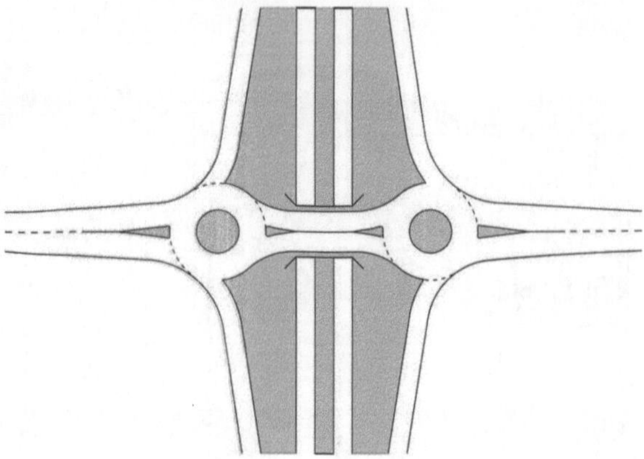

Fig. 2.41 Dumb-bell roundabout; sketch

accordingly find approaching ramps difficult. At a dumb-bell roundabout, speeds are significantly lower, as two roundabouts are a measure for traffic calming. Importantly, this type of roundabout has a low number of conflict spots (Fig. 2.42). At a "standard diamond interchange", drivers might make a mistake and turn towards the wrong direction at the ramp. At a dumb-bell roundabout, such an option is significantly lower. A dumb-bell roundabout even provides the possibility of completely eliminating the option of driving in the wrong direction—using the adequate deflection of a ramp. This configuration also allows for easy U-turns.

This type of roundabout is very common in different European countries (Figs. 2.43 and 2.44) and elsewhere. It seems that the more numerous dumb-bell roundabouts are located on the Canary Islands, where virtually all ramp intersections are constructed as a dumb-bell roundabout.

This type of roundabout is also becoming increasingly common in the USA. Examples of dumb-bell roundabouts are located mainly in Minnesota, Arizona, California, Indiana and in some other states.

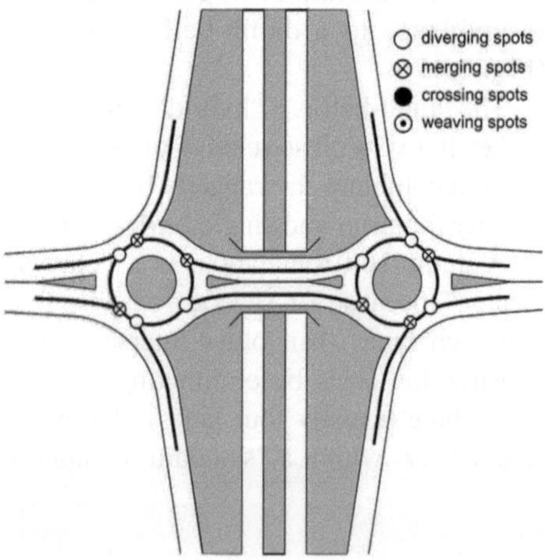

Fig. 2.42 Conflict spots on a dumb-bell roundabout

Fig. 2.43 Dumb-bell roundabout on motorway; The Netherlands

Fig. 2.44 Dumb-bell roundabouts on motorway; Slovenia

The main disadvantage of this type of roundabout is lower capacity than at the roundabout interchange with two roundabouts working less skillfully than one. The second disadvantage is that it is difficult to build this type of roundabout where a large roundabout has been built prior to the new one.

2.3.8 Ring Junction

Ring junction (or chain roundabout or magic roundabout) is, as is known to the author of this book, known only in the United Kingdom, and even there, there are only a few examples.

In general, this type of roundabout (Fig. 2.45) is located at a junction of more than four roads and consists of a two-way road around the central island with a few mini-roundabouts where it meets the incoming roads. The basic characteristic of this type of roundabout is that traffic may proceed around the main roundabout either clockwise (in the UK) via the outer lanes, or anticlockwise using the inner lanes next to the central island. The inscribed circle diameter is about 60 m. At each mini-roundabout the usual clockwise flow applies (in the UK).

A "ring junction" was formally defined for the first time in the TM.H2/75 [13]. It was defined as "an arrangement where the usual clockwise one-way circulation of vehicles around a large island is replaced by two-way circulation with three-arm mini-roundabouts and/or traffic signals at the junction of each approach arm with the circulating carriageway". The guidelines also state that ring junctions have been found to work well in solving problems at existing large roundabouts and that the conversion to a ring junction is an effective solution for very large roundabouts which exhibit entry problems. A ring junction will not operate successfully unless the signing is clear, concise and unambiguous [1].

It seems that the world's best known ring junction is in Swindon in Wiltshire, known as "Magic Roundabout". Its name comes from "The Magic Roundabout",

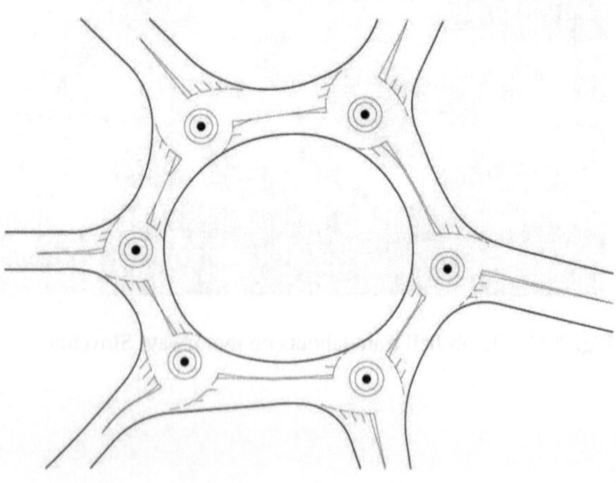

Fig. 2.45 Ring junction—left-hand driving; sketch

in the original French as "Le Manège enchanté", a French-British children's television program, created in France. The Magic Roundabout was constructed in 1972, according to the design of Frank Blackmore, of the British Transport and Road Research Laboratory. The solution consists of five mini-roundabouts arranged around a sixth central, anti-clockwise roundabout. Traffic may proceed around the main roundabout either clockwise via the outer lanes, or anticlockwise using the inner lanes next to the central island.

When the roundabout complex was first opened, the mini-roundabouts were not permanently marked out and could be reconfigured while the layout was finely tuned. A police officer was stationed at each mini roundabout during this pilot phase to oversee how drivers coped with the unique arrangement. In 2005, it was voted the worst roundabout in a survey by a UK insurance company, and in 2009 it was voted the fourth scariest junction in the UK [14]. However, the roundabout provides a better throughput of traffic than other designs and has an excellent safety record, since traffic moves too slowly to do serious damage in the event of a collision [15].

Similar systems (with five or six mini-roundabouts) can be found in various places in the UK (Colchester, Hemel Hempstead, High Wycombe…).

So, as previously stated, roundabouts in different countries differ in their layouts, and there is no "only one truth". A certain solution which is safe in one country could be very dangerous in another, and verbatim copying of foreign results could be dangerous and can lead to effects that are completely opposite than expected.

2.3.9 Roundabout with a Transitional Central Island

The importance of the central island of a roundabout has been extremely high from the very beginning of roundabouts' developments. In 1929 Watson criticized the decision of the London Traffic Committee for favoring squares or diamond shapes. These tended to increase the approach and entering speeds and also slow down the speed of rotation. It ensured that entrance would begin to take priority, and with the increasing general speed and volume of traffic, frequent "locking" would eventually occur. Watson's suggestions for overcoming the committee's "architectural objections" to circular islands included partial, split or double roundabouts [1].

A roundabout with a transitional central island is usually called a "hamburger roundabout" (the name came from the aerial view: the two halves of the central island look like the "bread", and the splitter island between two roads represents the "meat") but the terminus "split-roundabout", and "through-about", and "cut-through" roundabout are also in use (Fig. 2.46). The hamburger roundabout is a type of roundabout with a straight-through section of carriageway regarding major roads. It has a split central island with a splitter island between the two halves of the central island. The width of the intermediate splitter island is equal to the

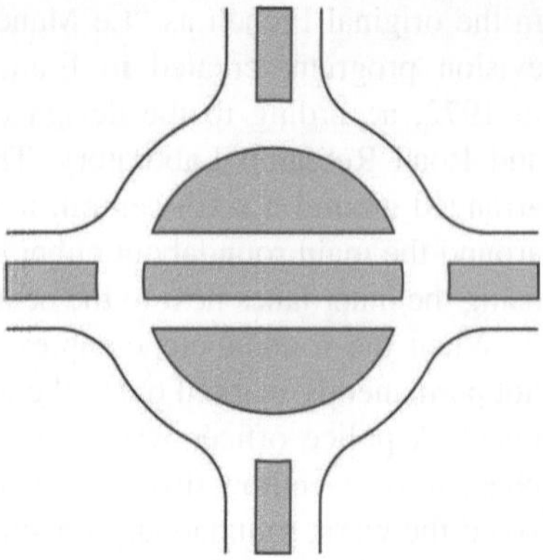

Fig. 2.46 Hamburger roundabout; sketch

length of one heavy vehicle or one bus (or more, but not less). The inscribed circular diameter of the hamburger roundabout is about 60 m or more.

It could be constructed as a one- or two-level roundabout. There are few variations of this type of one-level solution. One of them, in frequent use in the Canary Islands (Fig. 2.47) also includes splitter islands on approaches (Fig. 2.48) and for right turners (Fig. 2.49).

In the UK and Ireland this type of roundabout is still in frequent use, and is also very common in Spain and Portugal. This type of at-grade hamburger roundabout is often traffic signal controlled because of the large number of conflict spots (Fig. 2.50), and always lighting.

There is also a variation of hamburger roundabout on two-levels (Fig. 2.51). The main carriageway goes straight through the middle of the junction at one level (under or overpass), with short ramps connecting it to the roundabout at other levels. This variation of a hamburger roundabout is always traffic signal controlled.

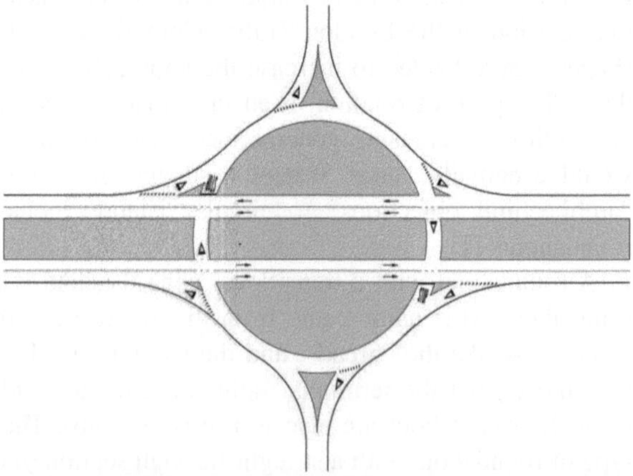

Fig. 2.47 Hamburger roundabout with splitter islands for right turners; sketch

Fig. 2.48 Stop and yield signs on the crossing with a straight-through section of carriageway

Fig. 2.49 Yield at the entrance of circulatory carriageway

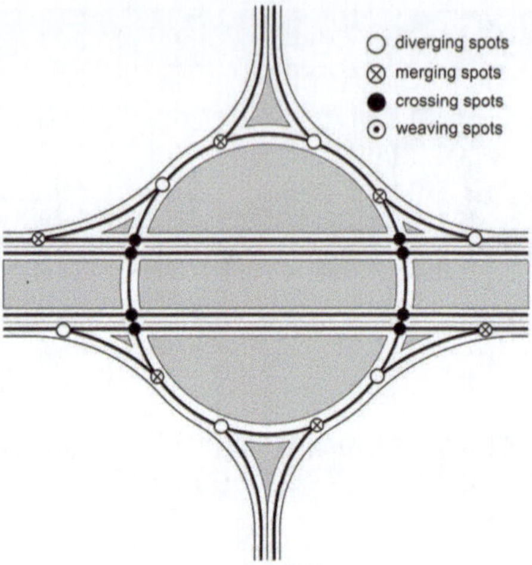

Fig. 2.50 Conflict spots on an at-grade hamburger roundabout

Fig. 2.51 Two-level hamburger roundabout; Barcelona, Spain

2.3.10 Roundabout with Segregated Right-Hand Turning Lanes (Slip-Lanes)

Standard one- or two-lane roundabouts (also some alternative types of roundabouts) sometimes incorporate segregated lanes for right-hand turners ("slip-lanes" or

Fig. 2.52 Segregated right-hand turning lane; Maribor, Slovenia

"segregated right-turning lanes" or in some European countries called "bypasses" or "free-flow lanes" or a "channelized turn lanes"). A slip-lane is a separate right-turning lane (Fig. 2.52) that lies adjacent to a roundabout, and allows right-turning movements to bypass the roundabout itself. A slip-lane provides traffic relief by allowing right-turning traffic to bypass the roundabout instead of passing through. A slip-lane is not a dedicated right-turn lane within a roundabout approach. The purpose of a slip-lane is to separate the flow of right-turning traffic, reduce delay and vehicle conflicts within the roundabout to improve capacity and safety.

The segregated right-turning lane is a recognized method in many countries (e.g., the UK, France, The Netherlands, Germany, the USA, Poland, Czech Republic, and Slovenia) of increasing capacity at a roundabout where a high proportion of the flow turns right. But, in some of these countries, guidelines advise that they can lead to speeding [16]. Therefore, it is necessary to stress that the designer needs to consider a number of factors, especially if vulnerable participants onto a roundabout are expected.

Two basic types of segregated lanes are known; non-physically segregated and physically segregated right-turning lanes. A non-physically segregated lane is a right-turning lane from a roundabout entry to the first (next) exit, separated from the roundabout entry, circulatory carriageway and exit by means of an island delineated using road markings only. A physically segregated right-turning lane is a right-turning lane from a roundabout entry to the first (next) exit, separated from the roundabout's entry, the circulatory carriageway, and the exit by means of a curbed island and associated road markings.

In both types, vehicles are channeled into the right-hand lane by road markings, supplemented by advanced directional signs. They proceed to the first (next) exit without having to give way to other vehicles entering into the roundabout.

Segregation by road markings is more common (drainage, snow plugging…) but it can be less safe as it can be subject to abuse by vehicles over-running the non-physical (painted) island.

Three different layout options for designing segregated right-turning lanes are known basically (Fig. 2.53), depending on the number of vehicles turning right-hand and on land availability:

- stop line at the roundabout's exit approach;
- yield line at the roundabout's exit approach;
- acceleration lane—a free-flow lane (Fig. 2.54).

From the capacity point of view (but not also from the traffic safety of non-motorized points of view) the best solution is with an independent lane for right-hand turning (Fig. 2.55).

Roundabout with independent lanes for right-hand turning is not an appropriate solution if vulnerable participants onto roundabout are expected. In that case, several other layouts' variations are possible, depending on the types of participants into a roundabout (Fig. 2.56). Layouts' variations differ from country to country because of local circumstances (human behavior and traffic culture).

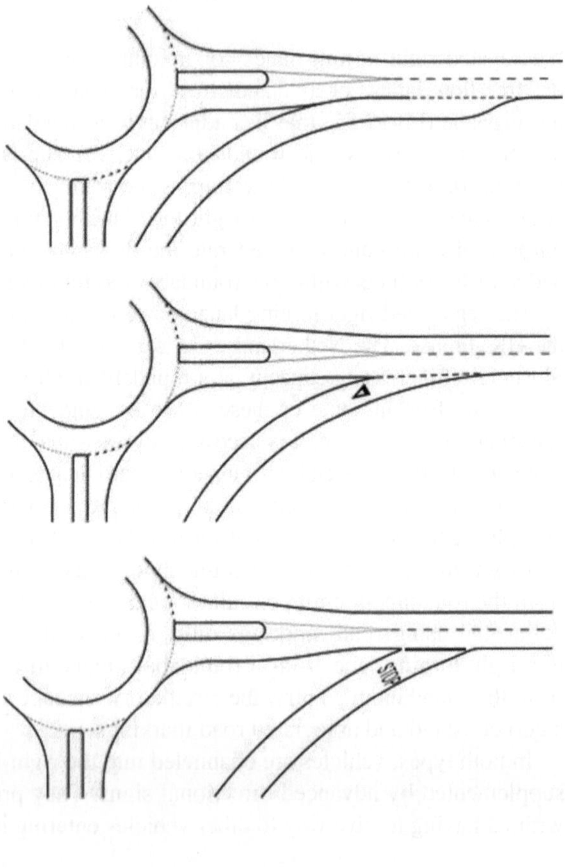

Fig. 2.53 Three different layout options for designing segregated right-hand turning lane

Fig. 2.54 Roundabout with four segregated right-hand turning acceleration lanes; City of Varaždin, Croatia

Fig. 2.55 An independent lane for right-hand turning

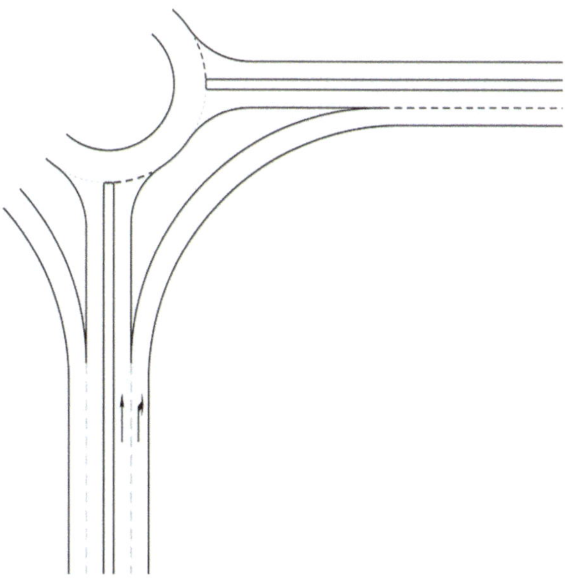

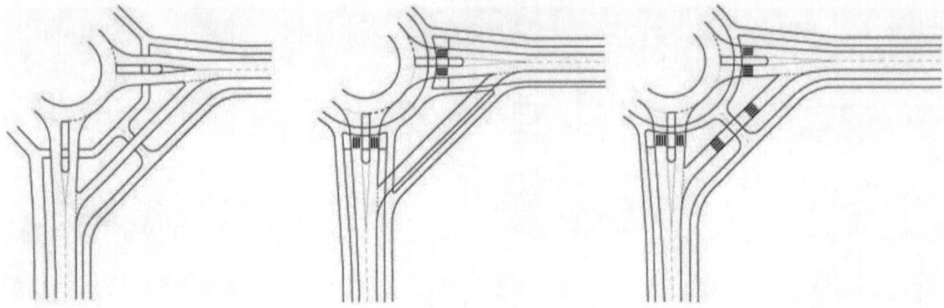

Fig. 2.56 Layouts' variations of segregated right-hand turning lanes if non-motorized participants are expected

The use of segregated right-turning lanes requires the designer to consider a number of factors (mostly traffic safety, capacity, and non-motorized participants) and should only be considered where its introduction would result in:

- an increase in the overall capacity of the entry or roundabout in question (compared to an alternative design);
- an improvement in the roundabout's safety (reduction of accident numbers or severity);
- provisions for pedestrians and cyclists.

The designer should determine whether facilities for non-motorized users are necessary, because segregated right-turning lanes can present particular difficulties for non-motorized users due to:

- possible high speeds of motorized vehicles;
- the extra widths of the carriageways at the entry and exit to be crossed;
- vehicle and non-motorized user conflicts due to large differences in speed at the pedestrian crossing;
- insufficient widths provided on pedestrian islands within physically—segregated right-turning lanes;
- confusion of vehicle flow direction due to the segregated nature of the right-turning lane.

As written above, if facilities for non-motorized users are necessary, the designer should determine whether they can be catered for adequately with a reasonable degree of safety and convenience within the roundabout design [16].

2.3.11 Signalized Traffic Circles

First, it is necessary to know that there are differences between signalized traffic circles and squares on the one hand and traffic signal controlled roundabouts on the other hand.

Signalized traffic circles and squares originated from the initial, old town squares with four or more intersecting roads. As written in Sect. 2.3.2, these old traffic circles and town squares are nearly always traffic signal controlled nowadays, because they are mainly located within city centers (usually with a lot of traffic). Traffic signals were initially installed on traffic circles and squares as part-time signals operating at peak periods, and this application is still common. The first experiment of traffic signals at a traffic circle in the UK was in 1959 [1].

Two main reasons exist for signalization of traffic circles and squares: entry flows were unreasonably balanced or old circulatory systems have been created as a result of multiple entry arms. Congestion was caused by tidal traffic conditions:

- high circulating speeds on large traffic circles or squares, which may make it difficult for other traffic to enter;
- when the major flow dominates the traffic circle or square to the extent that the remaining arms of the traffic circle or square experience severe difficulty;
- when a minor flow to the left of the major flow is dominant on the circulatory carriageway.

In these circumstances traffic signals have been installed at traffic circles and squares to counteract predictable operating imbalance by creating gaps in the circulating traffic.

The signalized roundabouts are a little bit different than signalized traffic circles. The signalized roundabouts originate from the UK and go back to the early seventies of the previous century; however, not until 1991 can we speak of their rapid expansion. From that year on, signalization became a popular method of traffic control in roundabouts and is now known also in the USA, Australia, Sweden, Ireland, The Netherlands, Germany, Belgium, Denmark, Turkey, Poland, and Slovenia. The traffic signal controlled roundabouts are discussed in Sect. 4.5.

References

1. Brown, M. (1995). *The Design of Roundabouts*. HMSO: TRL.
2. Highway Research Board. (1950). *Highway Capacity Manual*. USA, Washington DC: NAS-NRC.
3. Highway Research Board. (1965). *Highway Capacity Manual*. USA, Washington DC: NAS-NRC.
4. Žilionienė, D., Oginskas, R., & Petkevičius, K. (2010). Research, analysis and evaluation of roundabouts constructed in Lithuania. *The Baltic Journal of Road and Bridge Engineering*, *5*(4). doi:10.3846/bjrbe.2010.32.
5. Brilon, W. (2011). Studies on roundabouts in germany: lessons learned, In 3rd international conference on roundabouts, TRB, Carmel, Indiana, USA. Accessed May 25, 2011, from http://teachamerica.com/RAB11/RAB1122Brilon/player.html.
6. Tollazzi, T., Renčelj, M., & Turnšek, S. (2011). New type of roundabout: roundabout with "depressed" lanes for right turning—"flower roundabout". *Promet—Traffic & Transportation, Scientific Journal on Traffic and Transportation Research, 23*(5). doi:10.7307/ptt.v23i5.153.
7. Sawers, C. (2012). *Mini-roundabouts—Getting them Right*. A definitive guide for small and mini-roundabouts – right hand drive (update 2012), Moor Value Ltd.
8. Department for Transport. (2007). *Design of mini-roundabouts*. Design Manual for Roads and Bridges (DMRB), Vol. 6, Section 2, Part 2, TD 54/07, London, UK.

9. Kennedy, J. V. (2008). The UK standards for roundabouts and mini roundabouts. In National roundabout conference, TRB, Kansas City, Missouri, USA, 18–21 May 2008. Accessed May 25, 2008, from http://www.teachamerica.com/rab08/RAB08WWRKennedy/index.htm.
10. Kennedy, J. V., Hall, R. D., & Barnard, S. R. (1998). *Accidents at urban mini-roundabouts, Report TRL 281*. Crowthorne, Wokingham, GB: Transport Research Laboratory.
11. Brilon, W., & Bondzio, L. (1999). *Untersuchung von Mini-Kreisverkehrsplätzen*. Final report to the State Northrhine-Westphalia: Ruhr-University Bochum.
12. Schmotz, M. (2011). Capacity of mini-roundabouts. In 3rd international conference on roundabouts, TRB, Carmel, Indiana, USA, 18–20 May 2011. Accessed May 25, 2011, from http://teachamerica.com/RAB11/RAB1123Schmotz/player.html.
13. Department of the Environment. (1975). *Roundabout Design*. Technical Memorandum H 2/75, UK.
14. Disdale, J. (2007). *World's worst junction*. London: Auto Express Magazine.
15. Marshall, C. (2005). *The magic roundabout*, Chris's British Road Directory (CBRD).
16. Department for Transport. (2003). *Segregated left turn lanes and subsidiary deflection islands at roundabouts*, Design Manual for Roads and Bridges (DMRB), Vol. 6, Section 3, Part 5, TD 51/03, London, UK.

Chapter 3
Modern Roundabouts Design

3.1 Introduction

Analysis of literature shows that "modern roundabouts" nowadays exist in all European countries, as well as in more than 60 countries elsewhere in the world, so we could say that they are a world phenomenon.

Today in some European countries instead of the term "modern roundabout" they use "standard roundabout", which is understandable as the "modern" is already a long time "standard", especially in countries in which "modern roundabouts" have been implemented for several decades.

No uniform guidelines exist in Europe for the geometric design of roundabouts as specific circumstances (local customs, habits, traffic cultures...) differ from country to country. Certain solutions that are safe in one country could be dangerous in another. Consequently, most countries have their own guidelines for the geometric designs of roundabouts that are as far as possible adapted to their circumstances.

Roundabouts in different countries also differ in their dimensions and designs; the reasons being the different maximum dimensions of motor vehicles (mostly heavy vehicles) and special human behavior. As has been pointed out several times there is not "only one truth" in the case of roundabouts.

The most important questions regarding roundabouts' applications are:

- Is a roundabout an adequate solution under particular circumstances?
- Which type of roundabout is an optimal solution?

The point of this chapter is to try to answer these two questions. However, as stated before, it needs to be stressed that real circumstances differ from country to country.

T. Tollazzi, *Alternative Types of Roundabouts*, Springer Tracts on Transportation and Traffic 6, DOI 10.1007/978-3-319-09084-9_3

3.2 Criterion for the Acceptability of Roundabouts

Before we start designing a roundabout we must first check whether its implementation would be sensible under certain given conditions (specific location, specific traffic load, specific surroundings…). For this purpose we must evaluate the general criteria on acceptability for implementing roundabouts or the criteria for selecting an optimal intersection type. Evaluation of criteria for selecting an optimal intersection type is part of the study on acceptability regarding certain types of grade intersection.

A study on the acceptability of implementing individual types of graded intersections is a professional baseline for a concept and design project on the selected type of intersection.

The basic purpose of this study is an objective evaluation of acceptability regarding an individual type of intersection under given circumstances (traffic volume, space, surrounding, environment…), whereby the investor is guaranteed that the proposed solution is really an optimal solution, whilst on the other hand it protects the designer against unfounded requests for designing an unsuitable solution.

The most important part of the study when select an optimal type of intersection is the evaluation of global (general) criteria in order to justify the implementations of certain types of intersection. Different countries use different criteria to assess the acceptability of roundabouts [1–6]; however, the author presents a procedure, as given below, which is acceptable mainly for "standard" (one- or two-lane) roundabouts but could also be adapted for "alternative types of roundabouts" [7].

The following eight criteria are important:

- functional criterion;
- spatial criterion;
- capacity criterion;
- design (technical) criterion;
- traffic-safety criterion;
- front and rear criterion—criterion of mutual impact of consecutive intersections;
- economical criterion;
- environmental and aesthetic criterion.

3.2.1 Functional Criterion

When evaluating the functional criterion we are assessing the function (internal or transit traffic), role (whether we must increase the intersection capacity or to slow down the speed) and position (inside or outside an urban area) of the intersection in question within the total road network of a certain built-up area and which type of intersection is suitable for the anticipated future function of the intersection in question. This criterion is of a qualitative nature.

In short, we have to find an answer to the question: *What is the function or the primary role of the intersection in question?*

At first, we must establish whether we are dealing with an at-grade intersection or an up-grade junction. Results of many studies indicate that a one-lane roundabout is a good solution for the first intersection entering a built-up area. In these cases, a roundabout slows down speed, thus clearly warning drivers that they are entering an area of changing traffic conditions (lower permissible speed, non-motorized participants, pedestrians' and cyclists' crossings…).

On the other hand, a multiple consecutive traffic signal controlled "standard" intersection enables introduction of the "green wave" or synchronized regulation of the transit traffic through a sequence of intersections, which cannot be done at roundabouts.

Therefore, the introduction of multiple consecutive one-lane roundabouts on a arterial traffic route (where speed is important) is questionable.

Experiences of many countries also show that on heavily loaded urban arterial roads some other types of roundabouts (dumb-bell, dog-bone, and traffic signal controlled large roundabouts) present a good solution.

At the same time, we must also consider the vicinities and types of existing adjacent intersections, as it is well-known that unsuitable distances and different types of consecutive intersections lead to a negative mutual impact.

3.2.2 Spatial Criterion

When evaluating the spatial criterion we are establishing the availability of space for the intersection implementation. This criterion is of a quantitative nature.

In order to build a roundabout, we mostly need space for implementation of the inscribed circle diameter at the roundabout (Fig. 3.1). On the other hand, we need space for the implementation of separate lanes for left- and/or right-hand turns for the "standard" intersection.

We have to find an answer to the question: *Is there enough space available for the implementation of a properly dimensioned intersection?* [As the availability of land within an urban area (built-up surroundings, private property etc.) can often be questionable].

Usually, from the space criterion point of view, there is no difference between a four-arm one-lane roundabout and a "standard" four-arm intersection with separate lanes for left-hand turns. Also, from the space criterion point of view, dumb-bell and dog-bone roundabouts are inferior solutions to diamond interchanges.

Fig. 3.1 Verification of the spatial criterion for a roundabout on an existing Y intersection; City of Poreč, Croatia

3.2.3 Design (Technical) Criterion

From the design (technical) point of view, the implementation of a roundabout is appropriate and mostly recommended at the following intersections:

- the X, Y, A and K intersections (arms cross at a sharp angle);
- the F and H intersections (two consecutive T crossings positioned close together);
- with a large number of arms (five or more);
- where traffic signal controlled intersections are unjustified or infeasible.

This criterion is of a quantitative nature. The implementation of a "standard" intersection is suitable and recommendable for disallowed types of intersections (sharp-angled); in such cases the traffic regime can be regulated by traffic signs or traffic signals.

In short, we have to find an answer to the question: *What are the technical circumstances at the location of the intersection in question?*

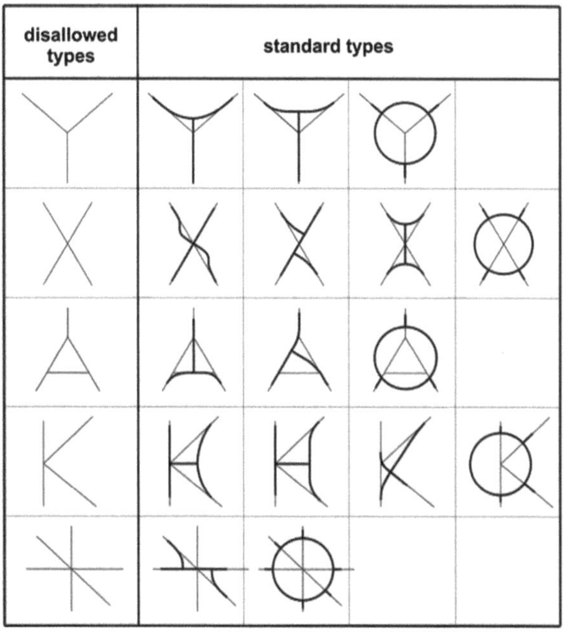

Fig. 3.2 Implementation of "standard" intersections at the disallowed (*sharp-angled*) types of intersections

On X, Y, A and K intersection (arms cross at a sharp angle), a one-lane roundabout is recommended but is not the only possible solution (Fig. 3.2) as it is possible to reconstruct the existing intersection into a "standard" three or four-arm intersection. In such cases, a roundabout solves the problem of the crossing angle and insufficient visibility.

For the reconstruction of an intersection in the form of F and H (two consecutive T intersections positioned closely together) a one-lane (or double-mini) roundabout is a good solution because it eliminates the mutual impact of two consecutive intersections positioned closely together.

At a multi-arm (five or more) intersection, a roundabout is a recommended solution (the other possibility is reconstruction into two three-arm T-shaped intersections). From the design point of view the implementation of a one-lane roundabout instead of the existing + or T intersection is unnecessary and unjustified. The implementation of a roundabout in these cases may be justified for the anticipated high speed at the new intersection or for the already known high speed at the existing intersection, the larger number of non-motorized traffic participants, slowing down speed at a major road, but this is the subject covered by the traffic-safety criterion discussed below.

3.2.4 Capacity Criterion

Verification of the capacity criterion means assessing a suitable solution in terms of quality regarding traffic flow transmission under given (today) and anticipated (future) circumstances and considering the volume of traffic flow at the end of a planned period. This criterion is of a quantitative nature.

We must find the answer to the question: *Will the proposed solution also comply with the anticipated volumes of traffic flow at the end of a planned period?*

When verifying this criterion we use different methods and approaches. A micro-simulation is the more popular at this moment, especially as there are many different low-cost programs on the market.

3.2.5 Traffic-Safety Criterion

The traffic-safety criterion is assessed regardless of whether we are dealing with a reconstruction or with a new intersection, since capacity and traffic-safety can be inversely proportional (in collision). This criterion could be of a quantitative or qualitative nature.

We must consider the following: *Will the selected solution under existing conditions and the conditions anticipated at the end of a planned period be a safe solution for all traffic participants?*

In the case of the existing intersection, we are comparing the level of traffic safety of the existing and anticipated solution; however in the case of an anticipated solution we are comparing several types of intersections (usually "standard" intersections, traffic signal controlled intersections and roundabouts). The comparison is made for all types of participants who are expected to be present at the intersection in question. Normally, the motorized and non-motorized traffic participants are processed separately.

Several international experiences show that small, one-lane roundabouts are appropriate solutions in front of schools, hospitals, and health centers. In contrast, large, standard two-lane roundabouts (which enable high speeds) are unwelcome at these locations.

Over the last few years, many different (new) types of roundabouts have become very popular in some countries, such as turbo-roundabouts, that could also be a good solution in these cases.

If we find out that the selected solution conflicts with some other global criteria, we should implement overpasses or underpasses.

3.2.6 Front and Rear Criterion—Criterion of Mutual Impact at Consecutive Intersections

We are considering the proximity and types of existing consecutive intersections because we know that unsuitable distances and different types of consecutive intersections lead to a negative mutual impact between them (gaps and delays). The more questionable are the alternating sequences of roundabouts and traffic-lighted intersections being close together.

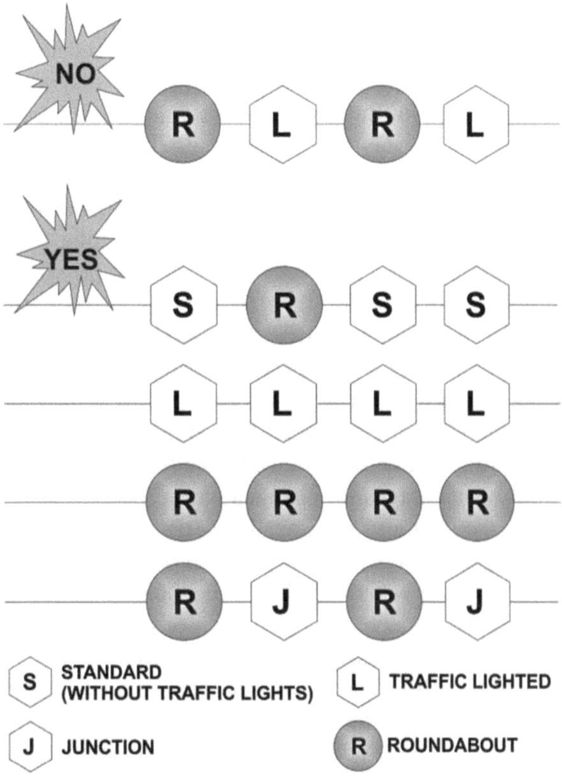

Fig. 3.3 Suitable and unsuitable combinations of different types of intersections [6]

This criterion could be of a quantitative or qualitative nature.

We must find the answer to the question: *How will our new intersection function in relation to other (existing) intersections?*

Today we know that some combinations of different types of intersections are suitable and others are not (Fig. 3.3), depending on distances between consecutive intersections and traffic volumes.

3.2.7 Environmental and Aesthetic Criteria

Negative impacts of traffic today are very important regarding the environment and the qualities of our surrounding life.

In regard to this criterion, we must assess how a certain type of intersection affects the qualities of surroundings and the quality of life (if we are dealing with a residential area). Thereby we mean the noise (when suddenly braking or accelerating), emissions from exhausts (gases and particular matters) when waiting in a queue or accelerating from a standstill position.

A traffic flow at a roundabout is determined by the amount of traffic on approach roads, consequently a well-balanced distribution over these roads ensures a good traffic flow. With a steady arrival of vehicles a roundabout can have shorter waiting time than a signalized junction. On intersections with unequal volumes, one-lane roundabouts (but not always on some of the alternative types of roundabouts) can have longer waiting time at certain access ramps. During the quiet hours, one-lane roundabouts have hardly any waiting time. It is impossible to give priority to a particular approach road by, for example, lengthening its green light time; which is possible at a signalized intersection. In general, the waiting time for cyclists and pedestrians is shorter on a roundabout, even without priority, than at a signalized intersection. According to several authors the exhaust emission increases by few percentages for CO and NO_X when a priority intersection is replaced by a one-lane roundabout, and if signalized intersection is replaced by a one-lane roundabout, the emission goes down by approximately thirty percentages for carbon monoxide CO and approximately twenty percentages for nitrogen oxide NO_X.

Firstly, we must analyze: *Which type of intersection would have minimum negative impacts on the environment (on the quality of life)?*

In this case we are dealing with a quantitative criterion as since some computer tools already enable the calculation of impacts on the environment from different types of intersections. Regarding the qualities of life of the surrounding residents, a roundabout is usually a better solution than a traffic signal controlled intersection, regardless of the weights ("performance index") selected during the calculation. There is less sudden braking and accelerating on a roundabout, less stopping, less waiting, less starting from a standstill position, and speeds are relatively low.

By aesthetic criterion we mean the compatibility of the selected solution with the surroundings' requirements, which is of a qualitative nature. Thereby we mean both the near and wider surroundings. The near surroundings comprise elements located within the immediate vicinity of the intersection in question. Namely, the near surroundings (e.g. surrounding buildings, trees, bridges …) may "deaden" the solution, obstruct visibility or obstruct proper implementation of access roads in a straight line (visibility).

We must also consider the wider surroundings (urban or rural), within which conditions at the intersections are located and the arrangement of its surroundings (avenues, green areas, construction of a central island, used materials and other). In this case, we are dealing with a quality criterion.

We are searching for answers to the question: *What should be the proposed type of intersection in the light of its surroundings and arrangement?*

Looking from the aesthetics point of view, roundabouts (and it doesn't matter which type) are usually better solutions compared to the "standard" intersections because a roundabout's central island represents a good spot for the (touristic) promotion of the town or region that we are entering. There are a lot of examples of nicely-designed roundabouts' central islands all over the world (Figs. 3.4, 3.5, 3.6 and 3.7), and in some countries really "everything is allowed" (but the visibility must never be in question).

Fig. 3.4 Place of the European bike week; Faaker See, Austria

Fig. 3.5 Promotion of the Provence, France

Fig. 3.6 Promotion of the Teotichuahan, Mexico

Fig. 3.7 Airport of Catania; Sicily, Italy

3.2.8 Economical Criterion

By economical criterion, we mean an economic justification for the proposed solution. We must seek answers to two questions: *What are the estimated expenses (implementation, maintenance) of the proposed solution compared to other alternative solutions?* and *How much will a community benefit because of less traffic accidents from some of the proposed solutions?*

Both criteria are of a quantitative nature.

The first verification deals with the expenses assessment regarding the implementation and maintenance for different types of intersections.

In general, we are comparing a "standard" intersection without traffic signals, traffic signal controlled intersection (electricity, changing of light bulbs, service and replacement of parts etc.) with a roundabout (lighting, maintenance of central and splitter islands, planting green areas etc.).

The second verification deals with comparisons between the expected numbers of traffic accidents per unit of transport load for different types of intersections. If we know the cost of an average traffic accident, we can assess a society's savings in the case of implementing a better and safer (and usually also more expensive) solution.

Now, we can consider two possibilities. Either we choose an intersection with the greatest number of positive characteristics, or we can first assign weights w_i to all the individual criterions C_i and choose an optimal type of intersection as that for which the weighted sum $\Sigma w_i C_i$ attains the maximal possible value.

3.3 Geometric Design Features

As was pointed out previously, roundabouts in different countries differ in their dimensions and designs; the reasons for this are the different maximum dimensions of motor vehicles (mostly heavy vehicles) and different human behavior (experiences). Consequently, most countries have their own guidelines for the geometric designing of roundabouts that are as far as possible adapted to their circumstances.

This chapter addresses the salient facts from some current different guidelines for the geometric designing of roundabouts. It is because of these differences that only those rules contained within a large number of guidelines are referred too. The texts of these guidelines have been truncated and slightly modified in some places, to avoid ambiguity.

In many guidelines, design procedures for roundabouts are based on:

- type of roundabout;
- typical (design) vehicle;
- inscribed circle diameter;
- circulatory carriageway;

- entry width;
- entry angle;
- entry radius;
- flare length;
- exit width;
- splitter island;
- entry and exit deflection;
- visibility;
- cross fall;
- traffic signs and road markings;
- lighting.

Some of the more important elements are presented in the following.

3.3.1 Type of a Roundabout

There are different types of roundabouts for different circumstances and conditions. The criteria for acceptability regarding some types of roundabout differ from country to country but usually these criteria are: location (urban or rural), expected capacity, space availability, expected traffic participants, and traffic safety.

It needs to be stressed that there are two broad regimes of a roundabout's operation. The first occurs in urban areas with high peak-flows, often with marked tidal variations and physical restrictions on space availability, and the second regime occurs in rural areas and is characterized by high approach speeds, low tidal variations and few physical constrains [7].

The classification of roundabouts differs from country to country; however the classification presented in Table 3.1 could also be useful.

Table 3.1 Types of roundabouts, their outer diameters and approximate capacities

Type of roundabouts	Outer diameter (m)	Approximate capacity (veh/day)
Mini urban	14–25	10,000
Small urban	22–35	15,000
Medium-sized urban	30–40	20,000
Large interurban	35–45	22,000
Turbo roundabout (medium-sized, urban and interurban)	40–70	40,000
Large interurban	>70	–

Note Approximate capacities are only approximate values for four-arm roundabouts with evenly-distributed traffic flows. The values given in the table are for information purposes only, and for solving specific cases. Each case should be checked in terms of actual traffic flows and the design and technical elements used

3.3.2 *Typical (Design) Vehicle*

The dimensional and dynamic characteristics of the "design vehicle" are the bases for selecting almost all the design elements of a roundabout (inscribed circle diameter, entry width, entry radius, truck apron …). As design vehicles' dimensions differ from country to country, consequently the values of the above-mentioned design elements also differ.

In several countries the "design vehicle" in the urban areas is the bus, and in rural areas a semi-trailer.

3.3.3 *Inscribed Circle Diameter*

The inscribed circle diameter is a very important design element in the cases of "standard" one-lane roundabouts and mini-roundabouts, but much less important in the cases of two-lane roundabouts (it doesn't matter which type). The inscribed circle diameter is the diameter of the largest circle that can be inscribed within the roundabout's outline. As pointed-out above, the inscribed circle diameter is a function of the design vehicle, which differs from country to country. In several countries, the inscribed circles' diameters of one-lane roundabouts vary from 26 to 28 m.

3.3.4 *Circulatory Carriageway*

The circulatory carriageway depends on the design vehicle (inscribed circle diameter), and is a function of the main role of a roundabout (speed reduction or higher capacity) and on the location of a roundabout (urban or rural area).

The circular carriageway should, if possible, be planned as circular. In the case of a "standard" one-lane roundabout, the truck apron is a part of the circulatory carriageway. In the case of larger roundabouts, there is no need for the truck apron.

3.3.5 *Entry Width*

The designing of roundabout entries is a complex procedure with several variables which must be addressed to ensure safe design and adequate capacity (entry width, flare length, entry angle, entry radius, and half-width approach carriageway). The entry width is an important feature that determines entry capacity. According to UK guidelines [3] it often needs to be larger in urban situations than in rural cases, but this is contrary to many of the other countries guidelines. Usually, entry width depends on the prescribed (or design) speed and on the typical (design) vehicle. It is also very important to know that vehicle deflection (deflection curve) is imposed on entry, because this governs the speed of vehicles through the roundabout.

In the UK, it is good practice to add at least one extra lane width to the lanes on the entry approach [1] because the relationship between entry width and capacity is quite significant. However, it needs to be stressed that in many countries this solution is now disallowed.

3.3.6 Entry Angle

The entry angle (Fig. 3.8) serves as a geometric proxy for the conflict angle between the entering and circulating traffic flows. Is some countries [1] there are differences regarding the entry angle constructions, depending on the distance between the offside of an entry and the next exit, but several countries [e.g. 2, 3, 8] have a rule that the entry angle needs to be from 30° to 40° [a compromise between capacity and (speeds) traffic safety].

3.3.7 Entry Radius

The entry radius is measured as the minimum radius of curvature of the nearside curb line at the entry. As the entry radius is a result of the design vehicle, its value differs from country to country, and is from 10 to 20 m. It needs to be stressed that larger entry radius means higher speeds on entry to the roundabout, and (in several countries) the entry radius must be smaller than the exit radius.

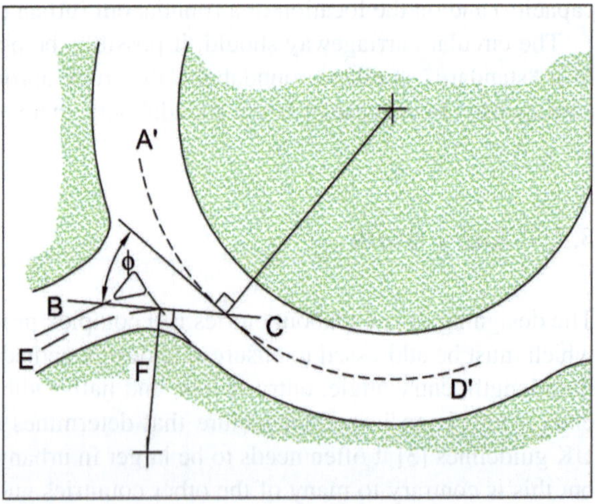

Fig. 3.8 The entry angle [2]

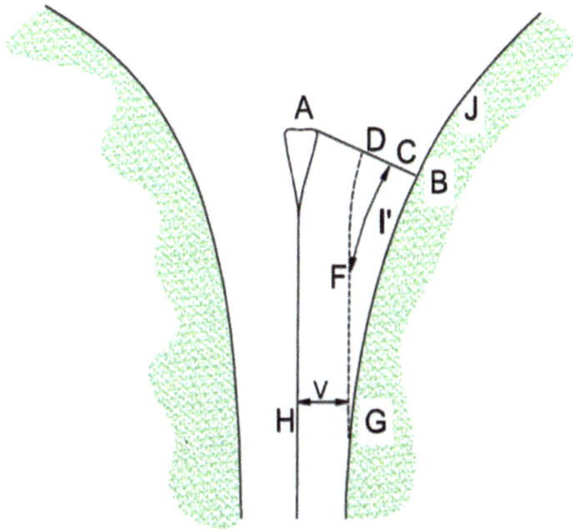

Fig. 3.9 The flare length [2]

It seems that in the cases of one-lane roundabouts, values between of 12 and 20 m are the more useful in several countries.

3.3.8 Flare Length

The entry path curvature is one of the more important determinants of traffic safety at roundabouts. The entry path curvature depend on entry width, layout of splitter island and flare length (Fig. 3.9), and is a measure of the amount of entry deflection to the right, imposed on vehicles at the entry to a roundabout. The "ideal value of flare length" differs from country to country, and is from a minimum of 10 m to a maximum of 40 m. A minimum length of about 10 m is desirable in urban areas, whilst a length of 40 m is considered adequate in rural areas. It needs to be stressed that flare lengths that are longer than 40 m have very little effect on increasing capacity, and in urban areas the use of long flare lengths is usually impossible due to land availability.

3.3.9 Splitter Island

The existence of a splitter island and its shape are very important for both capacity and traffic safety. A splitter island directs vehicles to the regular entrances and exits from a roundabout, and provides a higher level of traffic safety for pedestrians and cyclists crossing a roundabout's arm. The layout of the splitter island in several countries depends on the size of the roundabout (a triangle or a tear shape). On larger roundabouts, triangular-shaped splitter islands should be used, whilst tear-shaped islands should be constructed on small roundabouts.

3.3.10 *Visibility*

It needs to be stressed that we need to take into account different types of visibility (the forward visibility at entry, visibility to the left, circulatory visibility, and pedestrian crossing visibility). There are two main rules that need to be applied in terms of visibility at roundabouts in several countries. On:

- urban roundabouts, the driver may have visibility of the opposite exit from the roundabout, but this is unnecessary;
- suburban and rural roundabouts, the driver must be deprived of visibility regarding the opposite exit from the roundabout; it is achieved by landscaping or by a domed central island.

3.3.11 *Cross Fall of a Circulatory Carriageway*

Cross fall (and longitudinal gradient) provides the necessary slope that will drain surface water from approaches and the circulatory carriageway and super-elevation is arranged to assist vehicles when travelling around a curve.

Several countries use a rule that the standard cross fall for drainage on roundabouts should be 2 %, and should not exceed 2.5 %.

3.3.12 *Traffic Signs and Road Markings*

Traffic signs and road markings are highly important, especially at the start when implementing a new type of roundabout (e.g. mini roundabout or turbo-roundabout). But, traffic signs and road markings only influence traffic safety immediately after the roundabout is constructed, whilst later, their impact decreases or becomes void.

3.4 Effects of Layout Design Elements on Traffic Safety

An understanding of the geometric element's effect regarding roundabout layout on a driver's behavior or on accident potential, can lead to safer designs. However the variations and combinations of these elements, including such features as road signs, markings, lighting together with the total road environment, act in a complex way on a driver's perception. Therefore, in regard to the application of research results, the designer will always have to bear in mind the limitations of the data used and that all the factors are relevant to a particular site [8].

3.4.1 Traffic Safety of Motorized Road Users

In general, one-lane roundabouts are the safest type of at-grade intersection (having fewer vehicle collisions, fewer injuries, and fewer serious injuries and fatalities).

But at one- and two-lane roundabouts, some types of traffic accidents are possible that are not found at "standard" (three- or four-arm) intersections. It should be stressed that there are different traffic accident types at roundabout in different countries. The traffic accident types, shown in Fig. 3.10 are derived from the traffic maneuvers that can lead to traffic accidents.

The consequences of traffic accidents at a one-lane roundabout might also substantially differ from those at "standard" intersections. First of all, such

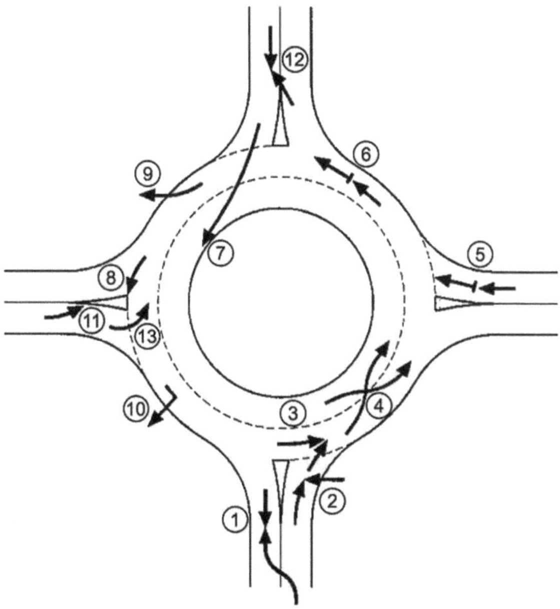

Fig. 3.10 Traffic accident types at roundabout

1. Overtaking before the roundabout
2. Collision with a pedestrian/cyclist
3. Collision at the entry
4. Collision at changing the running lane
5. Impact from the back at entry
6. Impact from the back at exit
7. Striking against the central island
8. Striking against the splitter island at entry
9. Slipping off the roundabout
10. Overturning
11. Striking against the splitter island at en-trance
12. Skidding (slipping) at exit
13. Driving to the opposite direction

consequences are less significant and, in principle, without fatalities and serious injuries. The reason lies in the fact that on a roundabout (and it doesn't matter which type), there are no frontal collisions with more serious consequences. On a roundabout, collisions between vehicles are mostly side-on collisions at acute angles or due to impacts from the back. Collisions amongst motor vehicles and cyclists (pedestrians) crossing the roundabout arm are the same as at "standard" intersection, whilst collision consequences are somehow less serious (reduced motor vehicle speeds at entry).

3.4.2 Traffic Safety of Cyclists

Traffic safety of cyclists at a roundabout depends primarily on the cycling traffic design method used within the roundabout's area, the designs of the splitter islands and on traffic signs and road markings. There are three types of cycling traffic design solution methods within the roundabout area worldwide:

- mixed motor and cycling traffic management (Fig. 3.11a);
- parallel cycling traffic management along the outer roundabout edge (Fig. 3.11b);
- separate cycling traffic management, parallel to curbs or in concentric circles (Fig. 3.11c).

Separate (independent) cycling traffic management at the roundabout is the safest management method. All intersections of motorized road users with cyclists (and pedestrians) are carried out perpendicularly, which provides an appropriate shape for the area of visibility. This solution also ensures that the only conflict points are located at crossings of the roundabout arms (Fig. 3.12), and even at these spots cyclists (and pedestrians) are (at least partially) protected by the islands.

Parallel cycling traffic management along the outer roundabout edge is less safe (with the exception of a roundabout with low traffic load) because the cyclist moves at the same level as the motorized road users. In order to increase

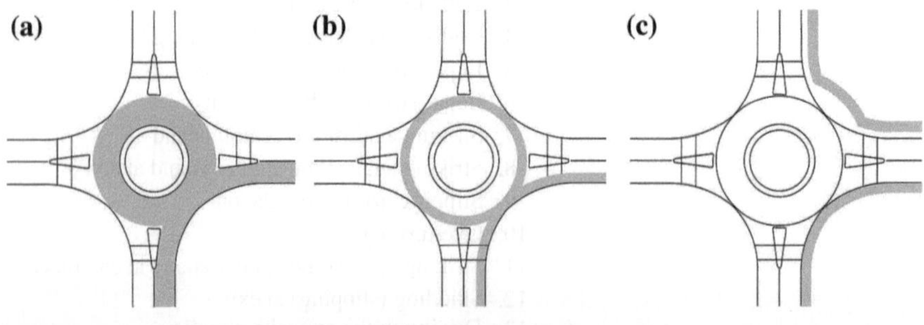

Fig. 3.11 Cycling traffic management methods at roundabout

Fig. 3.12 Separate—independent cycling traffic management within the roundabout area

the protection of cyclists in such cases, cycling areas are painted [in red: The Netherlands (Fig. 3.13), Germany, Belgium, Slovenia and some other countries, in blue: Denmark].

Mixed motor and cycling traffic management is the least appropriate method in terms of cycling traffic safety. In some countries (the UK, The Netherlands, Denmark, Sweden, and Belgium), this cycling management method was rather widely used in the past; however, following bad experiences (particularly in UK), this method of cycling traffic management is now completely abolished. The only exception includes small roundabouts in residential areas, in so-called "calm down traffic" areas in The Netherlands and in Germany (Fig. 3.14).

When introducing roundabouts in new surroundings, it is reasonable to implement some roundabouts with separate (independent) cycling traffic management at the beginning, and shift to roundabouts with parallel cycling traffic management over time. Such an opinion prevails in many countries. Consequently, designers in principle have chosen separate—independent cycling traffic management and this cycling traffic management method requires some additional space (and funds) but contributes significantly to increased traffic safety.

Later on, one of the three cycling traffic management methods will be selected based on the motorized traffic intensity and structure, cycling traffic flow intensity, and the roundabout's location in the overall road network settlement. It seems that the size of the cycling traffic flow intensity requiring the use of a specific cycling

Fig. 3.13 Parallel cycling traffic management along the outer roundabout edge; Maastricht, The Netherlands

Fig. 3.14 Mixed motor and cycling traffic management in Germany [10]

traffic management method is imprecisely defined by any regulations. Therefore, the decision is left to the consciences of the roundabout designers and investors.

Large roundabouts, especially those with faster traffic are unpopular with cyclists. This problem is usually considered at larger roundabouts by taking foot and bicycle traffic through a series of underpasses or overpasses or alternative routes.

3.4.3 *Traffic Safety of Pedestrians*

Pedestrian safety at a roundabout depends primarily on pedestrian crossings and visibility, and slightly less on the designs of splitter islands and traffic signs and road markings.

The availabilities of marked pedestrian crossings at a roundabout is necessary in order to provide sufficient traffic safety and convenience for pedestrians, on condition that they do not cause excessive congestion of motorized traffic. A pedestrian crossing will serve its purpose well only if it is located at a location where it attracts the greatest possible number of pedestrians (who would otherwise cross the road uncontrolled), and if it is sufficiently visible for drivers of motorized vehicles.

Pedestrians have to be able to perceive timely vehicles exiting or entering a roundabout. Special attention has to be paid to visibility for pedestrians in a roundabout combined with bus stops. Buses stopping at bus stops must not reduce the visibilities of pedestrians and/or drivers. The design of splitter islands has an impact on increased traffic safety for motorized and other road users. Consequently, providing pedestrian islands is recommended even in cases where all the conditions are not met (e.g. sufficient width).

Traffic signs and road markings only influence traffic safety immediately after the roundabout is constructed, whilst later, their impact decreases or becomes void.

3.4.4 *Measures—Conditions for a Safe Roundabout*

Global experience shows that there is a connection between some design and technical elements of a roundabout and the level of traffic safety. It is interesting to see that even in cases where design and technical elements are correctly selected, their combination might lead to a lower level of traffic safety.

Some results of comprehensive research from different countries are presented below. Results slightly differ but the common conclusion is that a safe roundabout should meet the following decisive conditions:

Directing of arms at a roundabout (Figs. 3.15 and 3.16) should be as perpendicular as possible (reduction of speed due to the deflection, appropriate shape of the area of visibility, etc.). The tangential direction of an approach lane at a roundabout results in misunderstanding of the right-of-way rule for vehicles within the

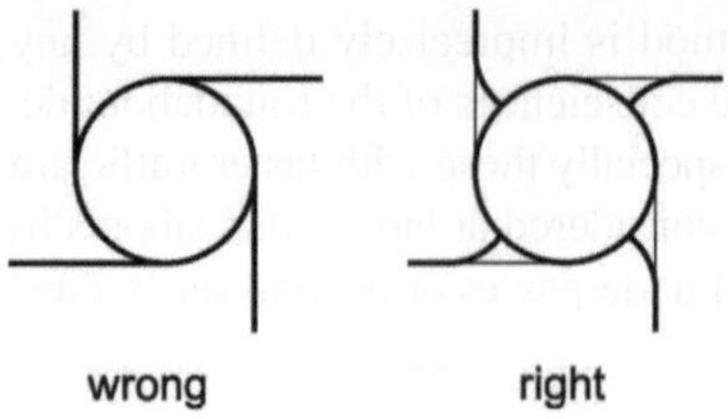

Fig. 3.15 Directing of arms at roundabout

Fig. 3.16 Right directing of arms at roundabout

circular course against vehicles at the entry, high speeds of vehicles at the entry, insufficient visibility for vehicles entering the roundabout and impacts of vehicles at the entry. The tangential direction when exiting from a roundabout requires an additional large steering-wheel turning angle and a large covered space at the entry.

Size of the entry radius; the speed at the roundabout entry directly depends on the size of the entry radius. Too large a radius allows excessive speeds at the entry, whilst too small a radius causes strikes against the central island or unwanted movement to the inner circular running lane in a two-lane roundabout;

Width of the roundabout entry and the length of an extension; the most dangerous traffic maneuver at a roundabout is when entering it, which is carried out within a relatively small area. Consequently, the entry shape is extremely important, both for traffic safety (undisturbed—constant driving at minimum speed, and waiting for empty space to enter the circulatory roadway) and the capacity (time gap);

Deflection of the driving curve through a roundabout is amongst the more important elements of traffic safety whilst driving through a roundabout. The curve has to be shaped as a double "S" curve encompassing three radii of harmonized sizes. A larger deflection of the driving curve causes lower driving speeds at the entry and exit, which results in higher traffic safety at a roundabout. Deflection of the curve can be influenced in two ways:

- through a change in the central island's size (which is better, but frequently infeasible);
- by the shape of a pedestrian island (which is worse, but frequently feasible).

Locations of pedestrian and cyclist crossings; it is reasonable to shift crossings outwards from a roundabout edge by one or two car lengths. This method allows for an increase in a roundabout's permeability at the same time, as pedestrians and cyclists obstruct the inclusion of vehicles into the circular current less intensively;

Splitter islands should be adjusted to the roundabout's size and to the speed foreseen at the roundabout (Fig. 3.17). At large roundabouts, using funnel-shaped islands is recommended, whilst at small roundabouts islands should be cone-shaped;

The truck apron at the central island (Fig. 3.18) is important for ensuring traffic safety on the circulatory carriageway. If there is no truck apron at the central island on the circulatory carriageway (large width of the circulatory carriageway), overtaking of circulating vehicles may occur leading to dangerous situations. The truck apron at the central island therefore represents a (visual) narrowing for small dimensional cars, and only exceptionally (for long vehicles) a useful part of the circular roadway. Regulations in numerous countries require a difference of 2–3 cm in height between the outer edge of the truck apron and the circular carriageway (discouraging cars from driving onto the truck apron of the central island, whilst this "tooth" does not represent any obstacle for long vehicles when driving);

Roundabout lighting determines a level of traffic safety at night. It is desirable to provide lighting at all arms of the roundabout and at the central island. It is desirable to place the lighting poles at the edge of the central island at large

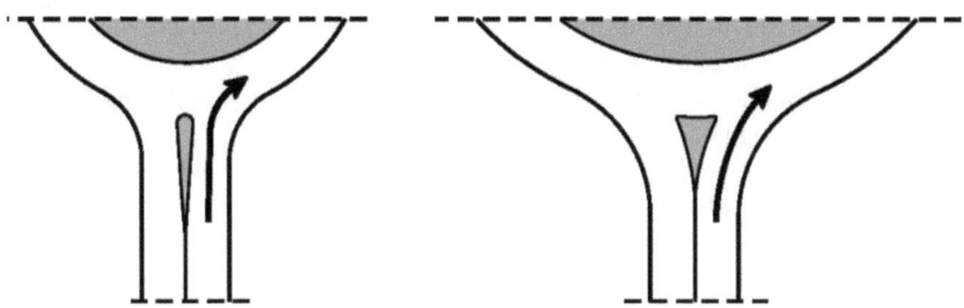

Fig. 3.17 Splitter island shape depending on the roundabout's size and the planned speed

Fig. 3.18 Truck apron at the central island

roundabouts, whilst it is sufficient to have the lighting placed only in the middle of the central island at small roundabouts (Fig. 3.19).

Arrangements at the central island (horticultural arrangement, fountains, monuments and other objects on the central island) are of significant importance for ensuring traffic safety at the roundabout by:

- shaping (domed) the land within a central island, it is possible to clearly warn vehicles that they are approaching a roundabout;
- partial hiding of vehicles on the opposite side of a roundabout, it is possible to eliminate negative driver feelings by a view of the traffic running around the whole roundabout (without the restriction of necessary visibility);
- plantations on the central island provide a good background for those road signs and signposting, placed at (and in front of) the central island;
- arrangements at the central island prevent being blinded by the headlights of oncoming vehicles.

In general, tree planting is unadvised on the central island, it might have an impact on the reduction of traffic safety for many reasons (lush crowns, cones, fruits, leaves…). The planting of trees is reasonable only at large roundabouts, and even there in such a manner as to primarily satisfy general visibility and the visibilities of traffic signs.

In urban areas, a central island's height (Fig. 3.20) is specified within several guidelines, but there also exist a lot of countries where this is left to the discretion of the designer. At sections outside urban areas where speeds are higher, the height of the central island has to prevent blinding by the headlights of oncoming vehicles at night; consequently, the height of a central islands summit is prescribed in many countries (1.1 m).

Fig. 3.19 The right way of roundabout lighting

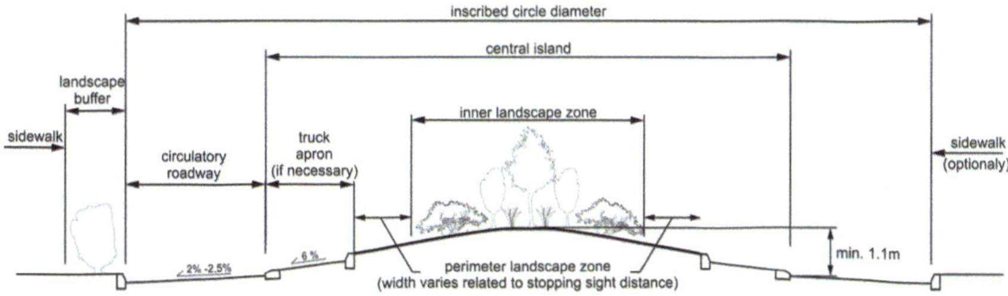

Fig. 3.20 A central island height and arrangement; according to Slovenian guidelines

3.5 Traffic Safety at Roundabouts—Some European Countries Experiences

As previously mentioned, analyze of literature show that nowadays “modern roundabouts” exist in all European countries, as well as in more than 60 countries elsewhere in the world.

Confidence in this now versatile form of road junction has been widely established in western (and also eastern) Europe by civil engineers, landscape planners, town traffic planners and most importantly—drivers. Many European countries (besides the UK), including France, The Netherlands, Germany, Belgium, Denmark as also Switzerland, are increasingly using roundabouts as an intersection design which bears comparison with, or is even more advantageous than other solutions.

At the start of this chapter, just briefly (because this is not the point of this chapter) overviews of some already known facts are presented.

Since the change to a give-way-on-entry rule at roundabouts ("giratoires") in France in 1983, the number installed has increased rapidly. According to some references, there are about 33,000 roundabouts in France at the moment. In France (Fig. 3.21), after the introduction of a roundabout, the frequency of accidents and fatalities reduced by more than 75 and 90 % respectively [8]. As a result roundabouts are being used more frequently and are recommended for junctions where the arms have relatively heavy balanced flows with high turning volumes. They are well-suited for the extremities of peripheral-urban by-passes. Although the change in the priority rule has improved safety at roundabouts generally, large (more than 50 m diameter) and/or non-circular with very wide or tangential entries appear to have lower safety performances.

Until about 1985 following successful French trials, no new roundabouts had been built in The Netherlands. In 1992 a "new wave" of roundabouts flooded The Netherlands and roundabouts started to be a popular solution for intersections and an estimated 300 or so new (off-side priority) roundabouts have been built. The

Fig. 3.21 Typical French two-lane roundabout, Aix en Provence

results of several research projects showed good safety levels for all kinds of road users and capacities of more than 2,000 veh/h. Until recently the new roundabouts have been relatively small and circular, with a single circulatory lane and narrow radial entry lanes. In many cases provision is made for heavy vehicles in the form of a raised strip 1.5–2 m wide at the outer edge of the central island. However the often large number of cyclists using them (a typically Dutch problem) has made it necessary to consider a suitable solution for accommodating these road users. Research into the safety of cyclists on new roundabouts has produced interesting results. Average speeds of 30–40 km/h were recorded within the vicinities of roundabouts compared to 50 km/h in the earlier situation. The numbers of conflicts between road users were the same as before but they were less serious. Observations of priority behavior showed that motorists still have more difficulty giving way to cycles and mopeds than to motor vehicles. Motorists within a dominant stream are less inclined to give-way. The use of direction indicators showed little consistency at roundabouts. Crossing times were similar to the previous situation but differences for the different types of movement were significantly reduced. The capacities of new roundabouts varied between 2,000 and 2,400 veh/h which was considered quite good for such limited dimensions [8].

Where practical and feasible, the Dutch have converted signalized intersections to roundabouts. A conversion to roundabouts is considered when serious signalized intersection accidents cannot be controlled by other means. They have successfully used one-lane and studies have shown a 60 % increase in intersection safety performance. The Dutch recognize that a one-lane roundabout design will not fit all situations. In order to achieve the best results, they have developed and implemented several different roundabout designs to address variable conditions. About 15 years ago they started with a growing interest on turbo-roundabouts and over the last couple of years with cyclist-friendly roundabouts. According to some reports, today there are about 200 turbo-roundabouts [9], three turbo-squares ("turbopleins") [10], and some new "cyclists' roundabouts" (Fig. 3.22).

Germany like other European countries has a long tradition of many decades regarding roundabouts. The roundabouts built between the 1930s and the 1960s were, however, limited in number and were mainly of a larger type with several lanes, both on the approaches and exits as well as on the circle. These traditional roundabouts have many similarities to those intersections which are called "rotaries" in the North-American context. Over the years, these conventional roundabouts gained a bad reputation regarding their safety and their limitations in capacity which—in spite of the large consumption in space—did not exceed an ADT beyond 40,000 veh/day [11]. Thus, they were no longer built after the 1960s and many of them were replaced by signalized intersections [12]. One of the reasons for this development is that in Germany all kinds of accidents (also damage-only accidents) are recorded by the police. The damage-only accidents are of frequent occurrence at these larger roundabouts, whereas accidents with personal injuries tend to be rather under-represented at the large roundabouts. With this kind of statistics the reputation of the larger roundabouts became worse over the years.

Fig. 3.22 "Cyclists' roundabout"; Eindhoven, The Netherlands

Interest in roundabouts in Germany is increasing and the number has grown rapidly since 1988. One-lane roundabouts of modern design have the best accident records of all the types of at-level intersections investigated. In particular the severity of accidents is quite low.

In regard to German regulations concerning vehicle size requirements, 26 m can be regarded as the lowest value for the inscribed circle diameter. However this leads to a very broad circulatory carriageway with a very small central island. Satisfactory roundabouts have been designed with an inscribed circle diameter of 30 m but in order to prevent high speeds through the roundabout and to provide good visibility the inner circle of the circulatory carriageway has a special surface bordered by a raised edge 2–3 cm high. Their experience [11] shows that this construction is only rarely used by the inner rear wheels of trucks and trailers, and it makes all car drivers and motorcyclists use the outer area of the circulatory carriageway.

Today, modern roundabouts in Germany include compact single-lane roundabouts (with diameters between 26 and 40 m), mini-roundabouts (with a diameters between 13 and 25 m), larger roundabouts (with a diameters between 40 and 60 m) with 2-lane access for cars and single-lane operation for trucks and turbo-roundabouts. It seems that German's specific one is the compact semi-two-lane circle (Fig. 3.23). The design of compact two-lane roundabouts is similar to the concept of one-lane roundabouts. The main difference is the width of the

Fig. 3.23 Typical German compact semi-two-lane roundabout; Oberhausen

circulatory carriageway. It is wide enough for passenger cars to drive side by side, if required. However, the circle lane has no lane marking. Large trucks and buses are forced to use the whole width of the circulatory carriageway making their way through the roundabout [11].

All these types have emerged as very successful regarding traffic safety as well as traffic performance, and on the other hand the traditional larger two-lane roundabouts have significant safety problems.

As was reported by Mike Brown [8], from the early 1980s roundabouts in Denmark have become more popular than traffic signal controlled intersections, but with rapidly increasing cycle traffic this has become a safety problem in towns. Accident analysis has shown that conflicts between approaching traffic and circulating cyclists has become dominant, causing 45 % of all accidents. Only 22 % of personal injury accidents affected drivers and passengers. The conclusion was that in Denmark roundabouts should not be "dynamically designed" as in the UK, but should be speed reducing and the carriageway width should be as narrow as possible. In order to reduce the speed of cars further, it is preferable to narrow the lanes by using rough "rumble" surfaces intended only for trailers (cut-in) in a strip nearest to the central island (inviting the car drivers to make a larger deflection around the central island than for goods vehicles). Splays at entrances provided only for the path needed for long vehicles, can be similarly surfaced [8]. And what is very interesting, their guidelines for urban roundabout design include the recommendation that a cycle track of 1.7 m minimum width is normally provided externally in the circulatory carriageway, with entry and exit connection to the cycle lanes of the radial arms (which differ in several countries).

What is written above is already "deja vu", but what is lesser known, very good experiences with roundabouts are also reported from Slovenia, Italy, Lithuania, Czech Republic, Croatia, the Former Yugoslav Republic of Macedonia and many other countries. The following text describes briefly some of the results published on the safety at roundabouts as experienced in some of these countries.

3.5.1 Slovenian Experiences

After a little more than twenty years since the first wave of roundabouts in Slovenia, there are currently more than 500 roundabouts [13]. Most of them are one-lane (Fig. 3.24) but there are also many two-lane, multi-lane, mini (Fig. 3.25), assembly one-lane, dumb-bell, traffic signal controlled, turbo-roundabouts (Fig. 3.26), and also assembly turbo-roundabouts installed all over the country [14].

When Slovenia became an independent country in the early 1990s, the need for establishing new legislation for the fields of road construction and road traffic appeared to be immense. Among many other measures, the Slovenian Transportation Ministry founded the Department for Roundabouts, the main task of which was the preparation of guidelines for the planning and constructing of roundabouts.

Fig. 3.24 One of the first Slovenian one-lane roundabout

Fig. 3.25 Typical Slovenian mini roundabout

Fig. 3.26 Slovenian turbo-roundabout

The process of introducing roundabouts into the Republic of Slovenia was spearheaded by a number of stakeholders. The key stakeholders included the traffic police, the media, and driving schools. The media especially played an important role by providing information to the largest number of users namely; drivers, pedestrians, as well as cyclists.

After the initial enthusiasm over the introduction of the first few roundabouts into Slovenia had subsided, questions concerning the justification of their installation and actual traffic safety surfaced. Considering that roundabouts in Slovenia were at the time a new phenomenon (with the exception of a few earlier examples), the concerns raised were completely understandable. Furthermore, there was no assurance that roundabouts in Slovenia would prove themselves appropriate, as they have abroad (in the UK, The Netherlands or Germany). However, this situation is common to all countries where "the best foreign guidelines" do not really exists, and each country must "do it their way".

The results of the Slovenian roundabouts traffic safety analysis were presented in 1997, 2004, and finally in 2010, with precise information on the number and consequences of traffic accidents. Based on traffic accident data at Slovenian roundabouts the major contributors were; traffic speed accounted for 63 % of all traffic accidents, incorrect movements of vehicles in which the drivers did not take necessary measures to ensure safe vehicular maneuvering (10.1 %), and inappropriate safety distances (7.9 %).

However, analysis shows that the percentage of cases involving excessive speeding has significantly decreased. This indicates that the drivers have received the message and have mastered the rules of driving at roundabouts which by design do not allow for excessive vehicular speeds. By design they ensure safety to all users whilst at the same time allowing a swift flow of traffic. A review of the results of the analyses showed that traffic safety significantly improved after the introduction of roundabouts and that the roundabouts in Slovenia have fulfilled their purposes and have, therefore, justified the expectations [15].

3.5.2 Italian Experiences

In Italy, as in other European countries, the process of roundabouts' implementation began in the 30s when motor vehicle traffic had increased; then it carried on until the 60s and then began again in the 80s. There had been an interruption for almost two decades but in the middle 90s roundabout building restarted and has gone on until today. There are no official numbers, but it is believed that in Italy since the 90s a few thousand modern roundabouts have been built [16].

During these years some meaningful evolutionary types, operating rules and fields of use have been recorded for their intersections. From the 60s until the beginning of the 80s, roundabouts with large diameters were mainly built (large roundabouts with many lanes for circulating roadways and arms) in suburban and urban areas as intersections between roads with high traffic volumes at

Fig. 3.27 Large roundabout in a rural area, Verona

the boundaries of metropolitan areas (Fig. 3.27). These intersections immediately appeared to be not only characterized by large land usage but also gradually unsafe in terms of the roads network usages. Extensive land usage followed instantly for the right priority rule existing at all intersections [16].

From the 60s to the beginning of the 80s, large squares in urban areas where it could be seen the coexistence among many urban functions and different transport modes, were transformed into roundabouts (Fig. 3.28).

Since the early 80s the above mentioned difficulties of large roundabouts in rural areas produced a progressive and quick rejection to building larger roundabouts (even if some of them were still being realized until the 90s), while in urban and metropolitan areas traffic lights at grade intersections become the more common solutions. The aim of the traffic lights' usage was also to provide new layouts of the roundabouts from a control point of view with greater diameters or of all circular organizations of the flow within urban and outside urban areas [16].

Starting from the 90s, following the example of what had occurred for over a decade in other European countries, the building of modern roundabouts, which spread rapidly throughout the country. The efficiencies of these types of roundabouts, as well as the reduction of the land-usage, were a direct consequence of the recalled new priority rule. This rule is also little by little being used for existing roundabouts with large diameters, thanks to specific redevelopment operations using road markings and traffic signs.

Fig. 3.28 Example of roundabout organization for a big Italian square, "Re Square", Roma

From 1990 till today extensive activity regarding the building of roundabouts has occurred without national guidelines. Functional evaluations (the assessment capacity and waiting times at legs) aren't usually made. However, a negative concept was also the spread among public administrations; who believed that the roundabouts were preferable in any case to other types of intersections, especially when these intersections were signalized. Road engineers, consultant engineers or public administration servants, proceeded in the planning and realization of these intersections by creative and absolutely subjective criteria, or at best, imitating foreign realizations (Figs. 3.29 and 3.30).

- their locations: at necessary and unnecessary locations;
- design elements: the number of lanes in circulatory carriageways vary from one to four, splitter islands have been designed with very different forms, with and without pedestrians' and cyclists' crossings, relationships between the entry, inner, and exit radii are very different;
- the construction details: the slopes of central islands, the presence or absence of monument elements on the central islands, the cross falls of circulatory carriageways etc.

Fig. 3.29 Actual example of Italian one-lane roundabout, Trieste

Fig. 3.30 Actual example of Italian compact roundabout

Finally, in July 2006, the Italian "Functional and geometric Standard for building road intersections" [17] was published. This standard also contains "Roundabout Intersections", specific information about the circular intersections, even if it's very concise. The wording is limited to just three pages with one table and three figures. It needs to be stressed that the Italian standard doesn't provide instructions on the geometric dimensions of a scheme or of its elements, nor when it can or has to be adopted. Therefore, it is clear that important matters like the geometric sizes of elements and vertical-horizontal alignments are not covered by the Italian Standard but depends on the planner's personal choices.

Three different types of roundabouts are defined within this standard (Table 3.2), according to the inscribed circles' diameters:

- conventional roundabouts with inscribed circles' diameters between 40 and 50 m;
- compact roundabouts with inscribed circles' diameters between 25 and 40 m;
- mini-roundabouts with inscribed circles' diameters between 14 and 25 m.

It can be seen that the Italian nomenclature does not agree with any of the foreign ones herewith considered as regards to roundabout dimensions, except for mini roundabouts with some slight differences.

In Italy there are as yet no useful hints from the standard nor from simulation studies or observations on the operating schemes regarding the aggregated traffic volumes that different types of roundabouts built in Italy can accommodate. Namely, they do not have their own capacity formula for the entry arms of a roundabout. Besides, a lot of Italian technicians do not make functional evaluations according to the traffic volume, despite the latter being clearly mentioned by the standard. Finally, there are no workings or scheduled national research addressing the safety analyses of the circular intersections.

On the other hand, scientific studies from an engineering point of view concerning roundabouts are being carried out very actively in Italy within the academic field [16].

Further, a lot of technical books concerning the geometric and functional designs of roundabouts have been published in Italy which represents good outlines aimed at applications of the international technical literature about this type of intersection.

It seems that unconventional solutions ("two-geometry roundabouts") started to be very popular in Italy [18]. Two-geometry roundabouts have a circular central

Table 3.2 Comparison among the roundabouts' classifications according to the inscribed circles' diameters D_e

Type	Instructions according the value of inscribed circle diameter D_e (m)			
	Italian	German	Swiss	American (FHWA)
Mini roundabouts	14–25	13–24	14–24	13–25
Compact roundabouts	25–40	26–60	22–35	25–30
Conventional roundabouts	40–50			30–60
Large roundabouts		55–80	32–40	

island surrounded by an oval circulatory roadway. A two-geometry roundabout is defined when the shape of the external margin is different from that of the central island, e.g. the central island is circular and the external margin is elliptic. In other words, the circulating roadway width is not constant. The combination of inner circular and outer oval shapes allows a reduction in carriageway width on the longer section and enhancement of path deflection. The same geometric combination is useful for enlarging the carriageway width corresponding to lower radii of curvature where high-occupancy vehicles require additional turning space. Roundabouts with unconventional geometries are also investigated for better accommodating oversize/overweight large trucks on routes necessary to key industry and the economy [18].

Finally, we can say that in Italy all the conditions exist in order to reach steadily and pervasively good planning and building practices for roundabouts.

3.5.3 Croatian Experiences

Several examples of roundabouts were constructed in the Republic of Croatia prior to the 1990s. From the global perspective the profession did not consider them to be positive solutions so traffic experts did not usually approve them. The criticisms were usually centered on the too large diameters of roundabouts thus enabling high speeds. The poor traffic situation was additionally weakened by a growing number of traffic lanes and other design-technical factors contributing to decreased traffic safety. The basic reason for the construction of few roundabouts could probably be assigned to the common legislation practiced by former state, unsupportive of roundabouts. The second reason principally refers to the poor experiences of the prior roundabouts [19]. Within the Republic of Croatia the majority of such examples were located in Zagreb (Fig. 3.31).

Contemporary large roundabouts were measured by contemporary valid knowledge (the size of a roundabout depends upon the required length for intersecting), thus enabling exceptionally high speeds, both at roundabouts' entrances and within the roundabout's circulatory carriageway. Such large roundabouts turned out to be poor for the Republic of Croatia.

Due to the positive experiences obtained in Western Europe and neighboring countries where roundabouts had rapidly increased during previous decades, 20 years ago roundabouts were used more frequently in both design ideas and practices in Croatia. The first systematic approach to roundabouts' implementations began in 2002 when Croatian guidelines for designing and equipping roundabouts [20] were issued. The disadvantage of these guidelines was that they weren't mandatory for the process of planning and designing roundabouts but instead represented a recommendation, irrespective applied by most designers.

The Republic of Croatia is presently preparing an updated version of roundabouts' guidelines by summarizing all the previous scientific and professional experiences and results. These guidelines will be mandatory and encompass all roundabout types: one-lane, assembly, and mini-roundabouts, with a special

Fig. 3.31 The old roundabout, Zagreb, Croatia

section given to turbo-roundabouts. The developments of these guidelines will significantly contribute to a systematic approach to the processes of planning and implementing traffic solutions based on roundabouts in Croatia.

Some Croatian regions have bravely initiated the construction of roundabouts using a high number of reconstructions of the existing intersections into the roundabouts [21]. Neighboring countries have had quite an influence over the number of the constructed roundabouts in Croatia. Two of the more western Croatian regions, Istria and North-West Croatia have the largest number of built roundabouts.

Presently Croatia has more than 200 roundabouts, mostly located within cities or within their suburban parts. These statistics however does not include roundabouts at special areas (big shopping centers, industrial zones, bus stations etc.). There is a certain trend of increasing roundabouts, with an increased number of cities applying this solution for their traffic problems. The common feature of roundabouts constructed lately is their relatively small (Fig. 3.32) or medium size (Fig. 3.33), while the construction of large roundabouts remains exception.

The City of Poreč serves as a good practice of implemented roundabouts. This city has so far constructed 17 roundabouts and it plans to build additional 11 (7 of them presently being designed while 4 are planned) which certainly places the City of Poreč at the very top of Croatian cities according to the number of implemented modern roundabouts (Fig. 3.34).

Mini roundabouts haven't been widely applied in Croatia, since only several practical examples of implementation can be found. Still, the implemented traffic

Fig. 3.32 Small one-lane roundabout, Novi Vinodolski, Croatia

Fig. 3.33 Medium one-lane roundabout, Labin, Croatia

solutions have produced good results from the aspect of traffic safety and traffic flow thus probably leading to the future implementation of these specific types of roundabouts.

In contrast, the Republic of Croatia already experiences several good examples of assembled roundabouts (Fig. 3.35), which are usually implemented as

Fig. 3.34 One-lane roundabout, Poreč, Croatia

Fig. 3.35 Assembled roundabout, Tar-Vabriga, Croatia

temporary solutions that should prove their feasibilities as permanent traffic solutions and as a measure for improving the traffic safety in case of missing the alternative permanent traffic solutions from the aspect of roundabouts.

In the Republic of Croatia is not implemented any turbo-roundabout yet. There are some propositions for reconstructing existing two-lane roundabouts into turbo-roundabouts but these are still perceived as preliminary solutions. However, it can be claimed that turbo-roundabouts will shortly become a regular traffic solution for Croatia as well.

Several scientific studies performed in Croatia indicate that modern roundabouts, from the standpoint of traffic safety, are quite acceptable solutions [22]. The more common types of accidents recorded on intersections before their reconstruction were side crashes and driving at unsafe distance. The reconstruction of classic intersections into roundabouts has significantly decreased the numbers and severities of traffic accidents (without mortalities), without any negative impact on traffic flow and the levels of service.

Despite all the challenges that occurred when implementing the roundabouts (inexperience in implementation of such solutions, non-existence of regulations, unequal approaches to designing etc.), it can be concluded that presently the Republic of Croatia is implementing an increased number of roundabouts, as a result of wider experiences of implementation, thus establishing roads for future development. It can be reliably stated that the Republic of Croatia will continuously increase its number of roundabouts by accepting of the European and global directions regarding traffic solutions on roundabouts.

3.5.4 The Former Yugoslav Republic of Macedonia Experiences

In the Republic of Macedonia, roundabouts were actually not known before 1991, with only three roundabouts built in different cities. The first roundabout in urban area was built in Skopje during the fifties of the past century. The roundabout was in place until the catastrophic earthquake that hit the city in 1963 (Fig. 3.36).

Between 1991 and 2006, the number of roundabouts increased slowly. There were no rules set out for roundabouts in that period, and the field of design was covered by a very few traffic experts with no experience in roundabouts. In that period, driving at roundabouts was not considered even a topic in driving school manuals.

Such circumstances continued until 2006, when Macedonia slowly began to embrace positive experiences of its neighboring countries. As Macedonia lacked own regulations for roundabout design, they used Slovenian and Croatian rules designing their roundabouts. Macedonia at present still lacks its own regulation for roundabouts, but there are slightly modified foreign rules in use. The use of different foreign regulations resulted in circumstances where roundabouts slightly differed from each other (depending mainly on traffic signs used), which resulted in confusion among drivers.

Fig. 3.36 The first roundabout in urban area, Skopje, an old postcard

Fig. 3.37 Small roundabout, Skopje

Since 2006, the number of new roundabouts in Macedonia has been steadily growing. There are an increasing number of cities choosing roundabout solutions solving their traffic problems. As a rule, these roundabouts are small, one-lane roundabouts in urban areas (Fig. 3.37). In 2008, they started modernizing the city of Skopje; consequently, most of new roundabouts are located in this city. Large roundabouts (Fig. 3.38) are designed only exceptionally, mainly due to spatial problems.

During the last 6 years a trend in the number of roundabouts has grown rapidly. There were 80 roundabouts built across the country, and nearly three times more are under the construction or project documentation at development stages [23].

Reconstruction of standard intersections into one-lane roundabouts has resulted in a significant reduction in the number of traffic accidents and their consequences. In traffic accidents at one-lane roundabouts, no fatal accidents were recorded. However, there were significant problems recorded of large, multi-lane roundabouts. This problem was so urgent that they increased the number of learning contents associated with driving at roundabouts in the material for driving license examinations, at the same time increased penalties for non-compliance of rules at roundabouts.

Fig. 3.38 Large roundabout, Skopje

Fig. 3.39 The first turbo-roundabout in Macedonia

Until now, there are no mini-roundabouts or assembled roundabouts built in Macedonia, mostly due to lack of experience in the design, in an absence of their own regulations.

On the other hand, their first turbo-roundabout was constructed in Skopje (Fig. 3.39) in 2011, mainly as a result of a strong initiative of their traffic experts, being the first turbo-roundabout in South-Eastern Europe.

They have very positive experience with the introduction of this type of roundabouts. Police department has recorded a very few accidents, all of them without injuries. Since its construction until today, there were only six traffic accidents recorded at the turbo-roundabout.

A very positive view of the turbo-roundabout is shared also by drivers. Their most frequently expressed positive opinion was that the driver had its own lane always available and that there was no weaving.

Lack of appropriate regulations for roundabout planning and design was the main problem, and it still is. However, taking into account all the challenges that Macedonia faced introducing roundabouts (lack of experience, absence of regulations, no uniform approach in design, etc.), we can conclude that at present high quality design solutions for roundabouts are available, which may be a basis for the development of their own rules, and a guide for further development of roundabouts in Macedonia [23].

3.5.5 Lithuanian Experiences

Year 1998 officially marks the year when the first modern roundabout was built in Lithuania, a solution which significantly helped to improve safety on Lithuanian roads and became widely popular between drivers, engineers and majority of road users [24]. Since then more than 50 roundabouts (Fig. 3.40) focused on traffic safety were constructed in this Baltic state and more than half as much are already designed and wait for its turn to be constructed.

Before implementation of modern roundabouts in Lithuania normally big diameter roundabouts, so called rotaries, were built. These roundabouts were designed to assure big capacity, and were mainly focused on the comfort of drivers. Old roundabout carriageways, entries and exits were wide, usually consisted of two or three lanes. These factors caused chaotic driving in roundabouts and were not providing much needed capacity or safety. Moreover, pedestrians and cyclists were completely ignored by the design of these rotaries.

Many things have changed since Lithuania became an independent state in 1990. The borders opened and Lithuanian specialists were able to improve their expertise abroad. Knowledge about one-lane modern roundabouts, turbo and mini roundabouts was gathered and implemented in first pilot projects. The Netherlands and Germany were two countries that mostly influenced Lithuanian roundabout design guidelines. For example in Lithuania, same as in the Netherlands, roundabouts with two-lane carriageway are not built anymore. They are considered to

Fig. 3.40 One of the first Lithuanian one-lane roundabout, City of Prienai

be less safe than turbo-roundabouts and in situations where one-lane roundabout capacity is not enough turbo-roundabouts are preferred.

Together with the construction of first modern roundabouts new design standards were prepared and published. In 2001 short description of modern roundabout design was described in R 36-01 “Intersections” standard. Main elements and parameters of mini, one-lane and large roundabouts were standardized [25]. In 2010 R ISEP 10 “Recommendation on implementing safe engineering solutions” was published. In this document roundabouts were divided into rural and urban roundabouts, turbo-roundabouts, main elements and brief capacity-geometry dependence were also described [26]. And finally, in 2012 MN ZSP 12 “Guidelines for roundabout design” were released. This document is the continuation of previous documents with 15 years of national experience and detailed capacity analysis. MN ZSP 12 is the document in which first steps towards modern roundabouts solutions for Lithuanian conditions are described [27].

Following innovative steps of Dutch engineers, different configurations of turbo-roundabouts are designed in Lithuania and traffic modeling tools are applied when evaluating traffic quality parameters. With these tools in complex situations it is much easier to evaluate the efficiency of the designed intersection. First turbo-roundabout in Lithuania was built in 2011 (Fig. 3.41). Geometry of this roundabout is similar to Dutch turbo-roundabout but lanes are not separated with curbs due to the possible problems with snow plowing in winter. The lane separating area is colored in red and together with milled noise straps was planned to deter drivers from lane changes. In Lithuania in 2013 there were two operating turbo-roundabouts, and around 10 of them are designed and are planned to be built in

Fig. 3.41 First Lithuanian turbo-roundabout, close to the city of Radviliškis

near future. No traffic accidents with casualties happened at already built roundabouts. A couple of accidents with car damage happened mostly due to geometry flaws and the lack of lane separating curbs. Situations when red painted areas did not serve as expected were observed. Therefore, practice as it is used in Slovenia, with modified version of Dutch curbs, could also be useful for future Lithuanian turbo-roundabouts. Otherwise, turbo-roundabouts proved as a successful and efficient solution which also improves traffic safety on Lithuanian roads.

3.5.6 USA Experiences

There has been a great amount of progress in roundabout design and construction in the USA especially over the last 15 years. A European observer might think that the number of roundabouts in the USA has simply "exploded".

The USA has been relatively slow to embrace roundabouts when compared to Europe. Although the USA was home to many of the first rotary intersections in the world, traffic circles had fallen out of favor in the USA by the 1950s. Older traffic circles (Fig. 3.42), located primarily in the northeastern states,

Fig. 3.42 Columbus circle, New York City

experienced serious operational and safety problems, as post war American traffic rapidly grew and had a tendency to "lock-up" at higher volumes. The modern roundabout following different design principles from those of the old traffic circles has been notably less popular in the USA than abroad, in part because of the USA's experience with the traffic circles and rotaries built in the first half of the 20th century.

The big difference between older designs and modern roundabouts needs to be mentioned. The older designs were large circles, which means:

- large circles = long weaving distances and large land usage;
- large curves = higher speeds;
- higher speeds = lower capacity and more severe crashes.

As of mid-1997, there were fewer than 50 modern roundabouts in the USA, in contrast with more than 60,000 in the rest of the world. However, since 2000, there has been an emergence of the modern roundabout in many states in the USA. The strong interest expressed in this type of intersection over recent years is partly due to its success in many European and Middle East countries as also in Australia and New Zealand, where the modern roundabout has greatly influenced the practice of intersection design.

As of 2014 there are an estimated 4,000 modern roundabouts nationwide, according to Eugene Russell, Civil Engineering professor and chairman of a national roundabout committee for the Transportation Research Board. The safety benefits have caused several states to adopt formal or informal policies that roundabout must be considered and in some states, have a priority over traffic signals. Also, their FHWA recommends roundabouts, and not traffic signals.

According to Bill Baranowski, the president of Roundabouts USA, the leading roundabout states in order are: Washington, Maryland, Colorado, North Carolina, Wisconsin, California, New York, Utah, Kansas, Indiana, Louisiana and Arizona.

Due to their significant benefits, the rate of roundabout construction is increasing yearly as experience broadens at the state and municipal levels. Many of the roundabouts are also well landscaped, thus improving the image of the roadway network, but more importantly, the decor and landscaping within the roundabouts are essential to making the roundabouts safe.

Modern roundabouts have become a subject of great interest and attention of traffic planners and civil engineers in the USA (Fig. 3.43) over the last 15 years. This interest is partly based on the success of roundabouts especially in Europe, where intersection design practice has changed substantially as the result of the good performance of roundabouts and their acceptance by European drivers. Maybe one of the more important reasons for this growing interest for roundabouts in the USA could be safety benefits, measured in European countries, especially for one-lane roundabouts, both in urban and rural conditions [28].

Roundabout development was relatively slow in the USA until the Federal Highway Administration weighed in on the design and planning of roundabouts with "Roundabouts: an Informational Guide". The guide was presented in 1999, adopted in 2000, updated again in 2010, and covers various types of roundabouts [29].

Fig. 3.43 Typical USA one-lane roundabout; Scottsdale, Arizona

Roundabouts have proven to be substantially safer than traffic signal controlled intersections. This is primarily due to the slower speeds required, in the range of 15–25 mph. When a crash occurs, injuries at these low speeds are unlikely. Because of their significant crash reductions compared to traffic signals, the FHWA Office of Safety lists roundabouts as one of nine recommended safety counter measures, and the Insurance Institute for Highway Safety also strongly supports roundabouts.

Some concerns were raised regarding pedestrians at USA roundabouts, especially with regard to the absence of clear right-of-way control. This perceived problem is related to some degree to the belief by the general public that traffic signal controlled intersections provide the greatest safety to pedestrians. These concerns tend to disappear after the pedestrians have an opportunity to drive in a roundabout. Public opinion surveys show that the attitude of users is generally positive after the roundabout has been in operation.

For bicyclists, the preferred arrangement in the case of one-lane and low-speed roundabouts is to terminate bicycle lanes before they reach the roundabout and to allow bicycles to circulate in mixed traffic through the circle. For larger, multi-lane roundabouts, it is preferable to provide separate bike paths, or to provide for mixed bicycle/pedestrian paths, or reroute bicyclists.

It has been determined that USA roundabouts can have significant benefits in terms of safety, delays, and capacity. Another major new benefit is related to the aesthetic and urban design improvements resulting from the landscaping and sculptural elements in the central island from the promotional point of view.

Fig. 3.44 Typical USA mini roundabout; White Center, Washington

Fig. 3.45 Landscaped traffic calming circle (neighborhood traffic circle); Seattle, Washington

Fig. 3.46 Painted traffic calming circle (neighborhood traffic circle); Seattle, Washington

Fig. 3.47 2-lane by 2-lane roundabout; sketch

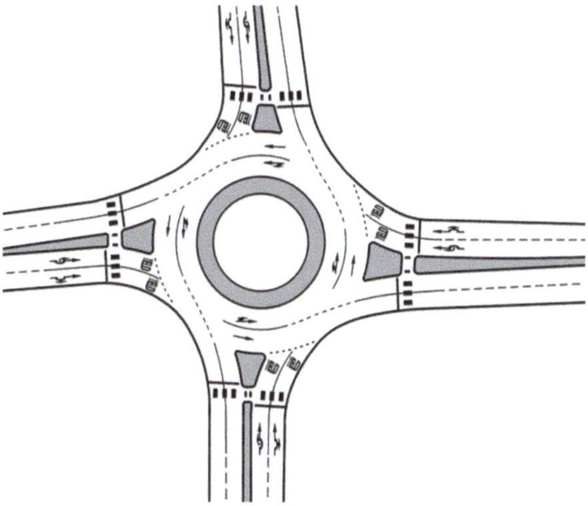

As it is known, at the moment, different types of roundabouts exist in the USA. Most of them are one-lane roundabouts but they also have a lot of mini roundabouts (Fig. 3.44), dog-bone roundabouts, traffic calming circles (Figs. 3.45 and 3.46), 2-lane by 2-lane roundabouts (with two lane entries from the major and minor street) (Figs. 3.47 and 3.48), dumb-bell roundabouts, and "major/minor roundabouts" (with two lane entries from the major street and one lane entries from the minor street) (Fig. 3.49).

Fig. 3.48 2-lane by 2-lane roundabout; Bellevue, Washington

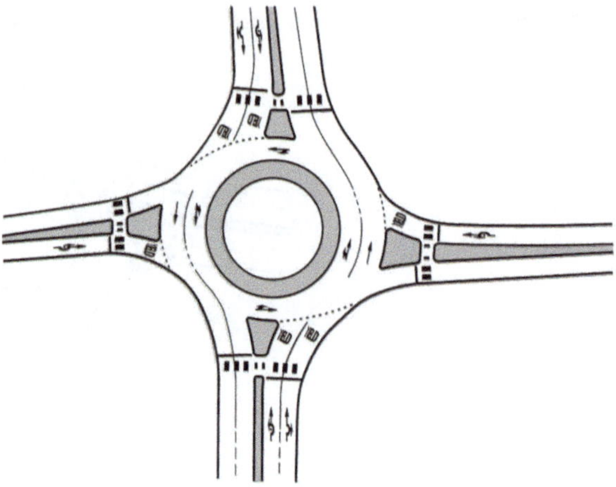

Fig. 3.49 "Major/minor roundabout"; sketch

Last couple of years City of Carmel, Indiana, has become the unofficial roundabout capital of the USA. The first roundabout in Carmel was constructed in 1997, instead of a four-way intersection with traffic lights. It was so successful that today eighty roundabouts are installed, and most remaining traffic signals will be replaced with roundabouts as the municipal budget allows.

3.5.7 Canadian Experiences

There has been a great amount of progress in roundabout construction also in Canada, where in 1999 the first roundabout was constructed. Also here a European observer might think that the number of roundabouts has simply "exploded".

According to Keith Boddy [30] at the moment of writing this book there are about 400 different roundabouts nationwide, which means one roundabout per 90,000 inhabitants (compared with e.g. the USA where there is one roundabout per 84,000 inhabitants), and the leading roundabout provinces are: Quebec, Alberta, British Columbia and Ontario. For Quebec it is also interesting that they have their own guidelines for roundabouts design.

They started in the same way as any other country; with beginning questions and challenges (reluctance with accepting roundabouts versus a traffic signal for intersection upgrades, and consultant understanding of how to design effective roundabouts).

Modern roundabouts have become a subject of great interest and attention of traffic planners and civil engineers in Canada especially over the last 15 years (Fig. 3.50).

Their strong interest is partly based on the success of roundabouts in Europe, where intersection design practice has changed substantially as the result of the good performance of roundabouts. Maybe one of the more important reasons for this growing interest for roundabouts in Canada could be (same as in the USA)

Fig. 3.50 Typical Canadian one-lane roundabout

safety benefits of roundabouts, measured in European countries. Due to their evidenced significant benefits, the rate of roundabout construction in Canada is increasing yearly as experience broadens at the provincial and municipal levels.

It has been determined that Canadian roundabouts can have significant benefits in terms of safety, delays, and capacity. Another major new benefit is related to the aesthetic and urban design improvements resulting from the landscaping and sculptural elements in the central island from a promotional point of view. Several Canadian roundabouts are well landscaped (Fig. 3.51), thus improving the image of the roadway network, but more importantly, the decor and landscaping within the roundabouts are essential to making the roundabouts safe.

As it is known, at the moment, different types of roundabouts exist in Canada. Many of them are one-lane roundabouts but they also have a lot of two-lane roundabouts, mini roundabouts, dog-bone and dumb-bell roundabouts, "major/minor roundabouts", and since last year also one turbo-roundabout (Figs. 3.52, 3.53 and 3.54) [31].

At the end it must be emphasized that new version of Canadian MUTCD [32] includes also roundabouts, that various jurisdictions are working on accessibility concerns to establish an array of tools for implementation when situations and context warrant, that the implementation of truncated domes in the Canadian context is being discussed and evaluated, and that their guidelines for roundabouts design are being developed.

Finally, we can say that also in Canada all the conditions exist in order to reach steadily and pervasively good planning and building practices for roundabouts.

Fig. 3.51 Well landscaped Canadian dumb-bell roundabout

Fig. 3.52 The first Canadian turbo-roundabout, approach; Victoria international airport, Victoria, British Columbia

Fig. 3.53 Circulatory carriageway at the first Canadian turbo-roundabout

Fig. 3.54 Splitter island and traffic signs

3.5.8 Japanese Experiences

Examinations of modern roundabouts began at the beginning of the 21th century in Japan. Even today some of old type circular intersections such as traffic circles or rotaries remain in small rural villages with very low traffic demands, and also there were several examples of roundabouts implemented on a trial basis. However the structures and/or operations of them were significantly different from those of modern roundabouts. Modern roundabouts are often misunderstood as rotaries, thus incorrectly recognized.

The research project on planning, design and application of modern roundabouts in Japan led by Prof. Hideki Nakamura, Nagoya University, Department of Civil Engineering, has been started from 2002. From 2006, this project has been extended to a voluntary research project of Japan Society of traffic engineers by involving researchers and practitioners, and their results were summarized as "Planning and Design Guide of Roundabouts in Japan (Draft)" of Japan Society of Traffic Engineers (JSTE) in 2009 [33]. In this guide, the application conditions and geometric design elements were investigated by referring to mainly German guidelines, as some of their design philosophy on safety and right-of-way conditions are also applicable for Japanese situations.

From 2009 to present, the project lead by Prof. Nakamura has been further supported by International Association of Traffic Safety Sciences (IATSS), and he has been continuously leading all of these projects for promoting implementation of roundabout in Japan.

Accordingly, a field operational test on roundabouts was conducted at Azumacho Rotary in Iida City, Nagano in 2010, by improving a five-leg rotary into modern roundabout. Such significant effects of safety improvements like speed reductions were demonstrated through this test and then the roundabouts got supported by a majority of users and residents of the vicinities. In accordance with the great success of this test, Iida City decided to reform the existing signalized intersection into a modern roundabout. In 2013, a brand-new five-leg modern roundabout was completed and started operation at Towa-cho, Iida City (Fig. 3.55), which is a leading example of implementing a roundabout by reforming an existing signalized intersection through removing traffic lights and reflecting the latest knowledge on modern roundabouts.

The breakdown of numerous traffic lights and the subsequent traffic chaos were caused by the unprecedented great East Japan earthquake in March 2011; meanwhile Japan has been faced with serious nationwide problems regarding the maintenance of over 200,000 traffic lights. From this painful experience, the Road Traffic Law was swiftly revised in 2013 so that roundabout can be legally defined as a form of intersection and priority is now given to the traffic on the circulatory roadway. This revision has been enacted from 2014.

Fig. 3.55 The first Japanese example of roundabout improved from signalized intersection; Iida City, 2013

Fig. 3.56 A six-leg roundabout improved from non-signalized intersection through a pilot study; Karuizawa Town

Under this circumstance, Road Bureau of the Ministry of Land, Infrastructure and Transport (MLIT) having jurisdiction over roads across the country subsidized field operational tests of roundabout at three locations in Japan (Fig. 3.56) from 2012 in order to verify its functionality and safety, and collect a variety of data.

All of the roundabouts implemented have so far been most favorably accepted by the local communities because of their safety performances and lower delays at intersections. Also, a large-scale roundabout experiment in Hokkaido was carried out by the National Institute for Land and Infrastructure Management (NILIM) in 2013.

All the results of these projects are intended to be fully utilized in the projected establishment of a roundabout guideline. In response to these changes of circumstances and activities by the government, it is expected that roundabouts will become more common throughout the country in the near future.

Although one of the specific features of roundabout examples in Japan is that application to multi-leg intersections where signalization was difficult, it is expected that standard four-leg roundabouts as a basic form will be popular in the future. With regard to the type of roundabout, so far the basic form will be a one-lane compact roundabout because of strict space restrictions and safety reasons. As it is not realistic to consider large-sized roundabouts in Japan, the regular inscribed circle diameter is considered to be 27–40 m. Major issues still to be investigated regarding roundabout applications are particularly the impacts and safety issues of pedestrians on capacity, as well as the safe treatments of cyclists.

References

1. Geometric design of roundabouts. (1993). *Design manual for roads and bridges* (DMRB) (Vol. 6, Sect. 2, Part 3). TD 16/93.
2. *TSC Krožna križišča* (Slovenian Guidelines for Roundabout Design). (2011). Direkcija Republike Slovenije za ceste.
3. Geometric design of Roundabouts. (2007). *Design manual for roads and bridges* (DMRB) (Vol. 6, Sect. 2, Part 3). TD 16/07.
4. Guide Suisse des Giratoires. (1991). Fonds de Securite Routiere, Institut des Transports et de Planification. Ecole Polytechnique Federale de Lausanne, Switzerland.
5. *Merkblatt für die Anlage von Kreisverkehrsplätzen.* (2006). Forschungsgesellschaft für Strassen- und Verkehrswesen (FGSV), Köln.
6. Benigar, M. (2008). *Mreža krožnih križišč v urbanem prostoru - temeljni principi definiranja,* 9. slovenski kongres o cestah in prometu, Zbornik kongresa, Portorož.
7. Tollazzi, T. (2000). *Sloveense rotondes: de eerste tien jaar ervaring*, Wegen.
8. Brown, M. (1995). *The design of roundabouts.* UK: TRL, HMSO.
9. Silva, A. B., Santos, S., & Gaspar, M. (2013). Turbo-roundabout use and design. In *CITTA 6th annual conference on planning research, responsive transports for smart mobility.* Coimbra, Portugal.
10. Werkgroep Evaluatie Geregelde Turbopleinen. (2008). *Toepassing Geregelde Turbopleinen*, Provincie Zuid-Holland.
11. Brilon, W. (2011). Studies on roundabouts in germany: Lessons learned. In *3rd International conference on roundabouts, TRB.* Carmel, Indiana, USA, May 18–20, 2011. Accessed May 25, 2011, from http://teachamerica.com/RAB11/RAB1122Brilon/player.html.
12. Stuwe, B. (1995). *Geschichte der Kreisverkehrsplaetze und ihrer Berechnungsverfahren* (History of roundabouts and their methods of calculation), Strassenverkehrstechnik, *39*(12).
13. Tollazzi, T., Renčelj, M., & Turnšek, S. (2011). Slovenian experiences with alternative types of roundabouts—"turbo" and "flower" roundabouts. In *The 8th international conference environmental engineering.* Vilnius, Lithuania: Vilnius Gediminas Technical University Press Technika, May 19–20, 2011.
14. Tollazzi, T., & Renčelj, M. (2012, February). *Slovenian roundabouts: We did it our way.* Roundabouts now (3rd ed., pp. 26–31). Accessed March 1, 2012, from http://digital.turn-page.com/i/58312?token=NmIwMzVkYmNiYzViMDYzZmU1YTM1ZTQ5MmE3OTc2Mjc1MDQ3N2NiMg.
15. Tollazzi, T., Renčelj, M., & Turnšek, S. (2011). Slovenian experiences with turbo-roundabouts. In *3rd International conference on roundabouts, TRB.* Carmel, Indiana, USA, May 18–20, 2011.
16. Mauro, R., & Russo, F. (2011). Roundabouts in Italy: A brief overview. In *International roundabout design and capacity seminar in connection with TRB 6th international symposium on highway capacity and quality of service.* Stockholm, Sweden.
17. Ministero delle infrastrutture e dei transporti. (2006). *Norme funzionali e geometriche per la costruzione delle intersezioni stradali,* (Functional and geometric Standards for roads intersections design).
18. Gazzarri, A., Pratelli, A., Souleyrette, R. R., & Russell, E. (2014). Unconventional roundabout geometries for large vehicles or space constraints. In *4th International conference on roundabouts, TRB.* Seattle, Washington, USA, April 16–18, 2014.
19. Tollazzi, T. (2007). Kružna raskrižja (Roundabouts), Monograph, IQ Plus. Kastav.

20. Hrvatske cete. (2001). *Smjernice za projektiranje i opremanje raskrižja kružnog oblika – rotora* (Croatian guidelines for designing and equipping roundabouts), Zagreb.
21. Deluka-Tibljaš, A., Babić, S., Cuculić, M., & Šurdonja, S. (2010). Possible reconstructions of intersections in urban areas by using roundabouts, road and rail infrastructure. In *Proceedings of the conference CETRA*. Opatija, Croatia, May 17–18, 2010.
22. Barišić, I. (2014). *Urbanistički parametri pri planiranju kružnih raskrižja,* (Urban Parameters for Roundabouts Planning), Doctoral thesis, Faculty of Architecture, Zagreb.
23. Hristoski, J. (2010). *Kružna raskrižja u Skoplju kao mera za otstranivanje opasnih mesta – crne tačke,* Zbornik radova sa Savetovanje za opasna mesta (crne tačke) na putevima u Republici Makedoniji i njihovo otstranjenje u funkciji bezbednosti prometa, Skopje.
24. Skrodenis, E., Vingrys, S., & Pashkevich, M. (2011). Lithuanian experience of implementation of roundabouts: the research of accidents, operation and efficiency. In *The 8th international conference environmental engineering.* Vilnius Lithuania: Vilnius Gediminas Technical University Press Technika, May 19–20, 2011.
25. Lithuanian Road Administration Under the Ministry of Transport and Communications. (2001). *Construction recommendations: Intersections.* R 36-01, Vilnius, Lithuania.
26. Lithuanian Road Administration Under the Ministry of Transport and Communications. (2010). *Recommendation on implementing safe engineering solutions.* R ISEP 10, Vilnius, Lithuania.
27. Lithuanian Road Administration Under the Ministry of Transport and Communications. (2012). *Guidelines for roundabout design.* MN ZSP 12, Vilnius, Lithuania.
28. NCHRP. (2007). *Roundabouts in the United States.* Report 572, Washington, DC, USA: Transportation Research Board.
29. NCHRP. (2010). *Roundabouts: An informational guid.* Report 672 (2nd ed.), Washington, DC, USA: Transportation Research Board.
30. Boddy, K. (2014). Canadian roundabouts - practice review. In *4th International conference on roundabouts, TRB*. Seattle, Washington, USA, April 16–18, 2014.
31. Murphy, T. (2014). The turbo roundabout, a first in Canada. In *4th International conference on roundabouts, TRB*. Seattle, Washington, USA, April 16–18, 2014.
32. Transportation Association of Canada. (2014). *Manual of uniform traffic control devices for Canada* (5th ed.). Canada.
33. Japan Society of Traffic Engineers. (2009). *Planning and design guide of roundabouts in japan*. Japan: Draft.

Chapter 4
Recent Alternative Types of Roundabouts

4.1 Introduction

Today, after many years of experience regarding roundabouts, there are still different ideas about the "ideal roundabout" with little consensus on the crucial effects of rules on how to negotiate intersections. The development of design rules and advice from an extensive body of research should allow civil and traffic engineers to produce the most effective forms of this junction type, even if for a variety of reasons this is not always carried out in practice.

It needs to be stressed that the roundabout intersection has been "at the development phase" since 1902, and this development is still in progress. One of the results of this progress is the several types of roundabouts in worldwide usage today called the "alternative types of roundabouts". Some of them are already in frequent use all over the world (hamburger, dumb-bell ...), and some of them are recent and have only been implemented within certain countries (turbo, dog-bone, compact semi-two-lane circle ...) or are still at the development phase (turbo-square, flower, target, with segregated left-turn slip-lanes...). Alternative types of roundabouts typically differ from "standard" one- or two-lane roundabouts in one or more design elements, as their purposes for implementation are also specific. The main reasons for their implementation are the particular disadvantages of "standard" one- or two-lane roundabouts regarding actual specific circumstances. Usually, these disadvantages are highlighted by low-levels of traffic safety or capacities.

4.2 Definition

Alternative types of roundabouts differ from standard (one- or two-lane) roundabouts in one or more design elements, whilst the purposes for their implementation are also specific.

T. Tollazzi, *Alternative Types of Roundabouts*, Springer Tracts on Transportation and Traffic 6, DOI 10.1007/978-3-319-09084-9_4

Alternative types of roundabouts should be implicated because of several reasons but mainly because of [1]:

- disadvantages of "standard" roundabouts in particular actual circumstances;
- changes of "actual circumstances" which in the past led to "standard" roundabout implementation.

In continuation some of them are presented in more detail, and, as has been pointed out several times, there is not "only one truth" in the case of roundabouts.

4.3 Assembly Roundabout

An assembly (temporary) roundabout is a temporary design solution usually placed within an existing "standard" three or four arm intersection and constructed with the uses of elements, traffic signs, road markings and equipment pursuant to traffic safety requirements, as intended for improving traffic capacity and/or traffic safety. Placement within an existing "standard" intersection implies the construction of a temporary roundabout, if possible within the boundaries of an existing intersection, i.e. "between the existing curbs" (Fig. 4.1). Construction of an assembly roundabout does not usually envisage considerable displacement of the existing intersection curbs, digging of asphalt, nor any other similar complex intervention in terms of finance and construction.

Fig. 4.1 An assembly roundabout—under construction; Maribor, Slovenia

An assembly roundabout must be constructed using elements, traffic signs, road markings, and equipment compliant with applicable regulations (guidelines) and safety requirements. It means, that the temporary solution must comprise the same elements used for a permanent solution (radii of proper size, central island, splitter islands, pedestrian crossings, traffic signs etc.), with a difference, that these elements are prefabricated and the traffic signs and road markings are temporary (in most European countries in yellow, in some in white color). As a rule, in several European countries the speed at temporary roundabouts is limited to 30 or 40 km/h.

An assembly roundabout could be constructed at three- or four-arm intersection, with or without traffic lights. If a temporary roundabout is constructed within the location of an existing traffic-lighted intersection, the traffic lights for pedestrians should be turned off, whilst the traffic lights for motorized participants should usually be set into the "yellow blinking (flashing) regime" (in several European countries), should be turned off (Fig. 4.2) or should be turned off and covered with a black blanket.

Different countries use different conditions and locations for appropriate use of assembly roundabouts, but these conditions and locations are not presented in their guidelines for roundabouts' designs, and are usually as follows:

Fig. 4.2 Turned off traffic lights at an assembly roundabout on an existing traffic lighted intersection; Jesolo–Punta Sabbioni, Italy

- temporarily changed traffic conditions (e.g. a temporary change in traffic flow in the major and minor traffic flows) during the summer tourist season, fairs, events etc.;
- temporarily heavy traffic flow (e.g. construction site at the intersection itself, construction of a fourth arm within an existing three-arm roundabout, etc.);
- during the process of proving the suitability of constructing a roundabout as a permanent solution;
- at the time of constructing a roundabout as a permanent solution (for the undisturbed functioning of the roundabout at the time of construction, i.e. for the protection of construction workers when constructing the central island as well as the inscribed circle diameter);
- as a measure taken for traffic calming at existing intersections that are not traffic-lighted (if the speeds on the major road are excessive, or if the vehicles on the minor road have problems when merging or diverging the main traffic flow);
- for the purpose of immediately facilitating bad traffic safety conditions (if there is currently a lack of financial means to construct a permanent solution).

A temporary design solution implies a period during which [1]:

- traffic conditions are changed;
- traffic flow is disturbed;
- experiments for proving the suitability of constructing a roundabout as a permanent solution are conducted;
- a roundabout is being constructed as a permanent solution;
- the period from the time of decision that the roundabout is a better solution than the existing solution, in terms of traffic safety or capacity, to the time of constructing the permanent solution.

All stages, i.e. processes prior to the construction of a temporary roundabout are identical to the procedure for constructing a roundabout as a permanent solution. The procedure comprises three stages:

- inspecting the justification (appropriateness of) for constructing a temporary roundabout;
- designing the temporary roundabout;
- constructing the temporary roundabout.

Inspecting the justification (appropriateness of) for constructing a temporary roundabout is exactly the same procedure as constructing a permanent solution with a roundabout. At this stage, the fulfillment of the general criteria is inspected in regard to the appropriateness of constructing a temporary roundabout as a temporary solution. In the event that an evaluation has already been carried out for a permanent roundabout solution, the inspection of the justification does not have to be performed.

The design procedure of an assembly roundabout is identical to the design procedure of a permanent solution, i.e. it is thus somewhat more complex, as the "human factor" should be taken into account:

- customization of users to the previous solution (driving "by heart");
- confusion of users regarding the double road markings (if the manager does not require the removal of old markings);
- necessary channeling of pedestrians (to prevent crossing roundabout arms outside of pedestrian crossings); as well as
- the requirement for unchanged curb line (inspection of the curve with the design vehicle for the most critical maneuver).

Only the construction of an assembly roundabout depends on the selection of temporary elements. An assembly roundabout can be constructed from:

- concrete prefabricated segment-shaped elements of various curvatures (for construction of a central island) (Fig. 4.3);
- prefabricated (rubber or steel) curbs (Fig. 4.4);
- plastic protective barriers (Fig. 4.5);
- or (sometimes) with any other prefabricated elements (Fig. 4.6).

Fig. 4.3 Central island of an assembly roundabout, constructed from concrete prefabricated segment-shaped elements, Genova, Italy

Fig. 4.4 An assembly roundabout constructed from prefabricated rubber curbs, Izola, Slovenia

Fig. 4.5 Assembly roundabout, constructed from plastic protective barriers, San Donà di Piave, Italy

Fig. 4.6 Central island of an assembly roundabout, constructed from concrete sewage pipes, Wülfrath, Germany [2]

At assembly roundabouts, splitter islands are made of the same elements as used for the central island. At temporary roundabouts, a traversable part of the central island (truck apron) is usually not constructed in a fixed manner, but is only marked with a horizontal full-line on the outer edge of the traversable part of a central island.

Calculating the capacity of an assembly roundabout is identical to calculating a permanent solution. In order to calculate the capacity of a temporary roundabout, data regarding the volume and structure of traffic flows (traffic counting) should be acquired, which—unlike at permanent constructions—takes a shorter planning period.

As a rule, in several countries no soft landscaping is carried out at the central island of a temporary roundabout, and no sculptures or other similar structures are placed there. The installation of panels, boards and other structures or devices for visual or audio information and advertising on the central island of a temporary roundabout is usually forbidden.

If the main elements of an assembly roundabout are prefabricated plastic protective barriers, they must be filled with water for reasons of elements' stabilities and traffic safety. If a temporary solution is on during winter, some kind of environmentally-friendly solution for preventing freezing should be added to the water in prefabricated elements.

No problems arise along the left edge (next to the central island) during the winter removal of snow from the circulatory carriageway, but some work is required along the right edge (outer edge of the roundabout) because snow must be removed manually.

An assembly roundabout could be constructed as a mini, "standard" one-lane, dog-bone or even as a turbo-roundabout (Figs. 4.7 and 4.8).

Fig. 4.7 Assembly turbo-roundabout, Koper, Slovenia

Fig. 4.8 Assembly turbo-roundabout, Koper, Slovenia

4.4 Traffic Calming Circle (Neighborhood Traffic Circle)

Traffic calming circles (in some countries called "neighborhood traffic circles" or "micro roundabouts") are a measure of traffic calming in residential areas [3–5]. A traffic calming circle is, in general, a small circular island constructed within an existing "standard" intersection in order to provide geometric control for slowing-down traffic. This type of roundabout could be built at sites with inadequate space for building a conventional "standard" one-lane roundabout. Traffic calming circles exist at these smaller junctions to avoid the use of signals, stop signs, or the necessity to give way in favor of one road. A traffic calming circle provides control of traffic flow because it physically requires all types of vehicles (also motorbikes) to slow down in order to maneuver around them (Fig. 4.9).

In several European countries this type of roundabout generally works in the same way as other roundabouts types in terms of "right-of-way". As they are yield-controlled (because they are a type of roundabout), they typically include raised channelization for guiding an approaching driver onto the circulatory carriageway, are usually equipped with pedestrian crossings, and left-hand turning movement for large vehicles is disallowed. But also here, there are some differences. In the USA, for example, at some traffic calming circles, left-hand turning movements for larger vehicles are allowed to occur in front of the central island.

Fig. 4.9 Traffic calming circle (neighborhood traffic circle); Seattle, Washington

From the European point of view it could be potentially conflicting with circulating traffic, even if it seems that this solution works well in the USA.

In general, only the painted circle is unpopular in European countries. In most of them, if there is limited space, the traffic calming circle can be driven over by cars—low domed central island or elevated platform, and where space is available the traffic calming circle may even have a landscaped traffic calming circle. Generally, it means two types of traffic calming circles exist.

In The Netherlands, for example, a specific type of neighborhood traffic circle, the so-called "punaise" ("pushpin" or "road stud type roundabout") is preferred (Figs. 4.10 and 4.11). This type of neighborhood traffic circle is constructed in such a way that between the outer edge of the central island and curbs is a space of width 1.5 m, for smooth cyclists' traffic [4].

As pointed out above, in European countries only the painted circle is, in general, unpopular. Several countries have reported that where a painted circle is used it is very important that the diameter of a central island is larger than the width of the approach road (deflection). The experiences of many European countries also show that the influence of a painted central island is very low (low speed reduction). The negative aspect of a painted traffic calming circle is also that traffic turning left may turn before reaching the traffic calming circle and thus cuts the corner.

The influence of a raised elevation (low domed or elevated platform) is slightly better (better speed reduction). In this case the outer radius of the traffic calming circle needs to be from 5 to 10 m, and the diameter of the central island equal to the width of an approach road. The good experiences of European countries show that the center of a central island should be increased from 12 to 14 cm. In detail: If the circle island is low domed, its diameter depends on the dimensions of the intersections (widths of approaching roads), and is in general from 2 to 4 m. The top of

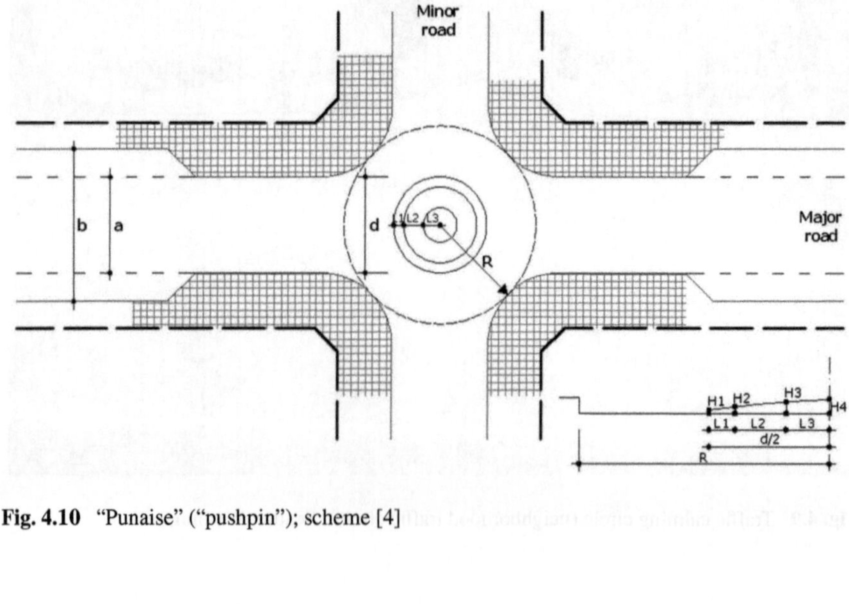

Fig. 4.10 "Punaise" ("pushpin"); scheme [4]

Fig. 4.11 Typical "punaise" ("pushpin"); Geleen, The Netherlands

the central island at its center also differs from country to country and is from 7 to 12 cm. In the case of a circular elevated platform the size of the diameter is equal to the widths of the approaching roads, and the top of the central island is usually equal or less than 14 cm. This type of traffic calming circle needs physical and visual support: trees on approach roads, different material on the pavement, and clear discernment ability through vertical elements and public street lighting. However, negative aspects of traffic calming circle with a raised elevation include the fact that traffic turning left may turn before reaching the traffic calming circle and thus cut the corner. A negative aspect of this type of traffic calming circles is also the increase in noise pollution and vibration by overrunning the central islands.

In general, the best experiences (high traffic safety level, good capacity) are with a landscaped traffic calming circle (Fig. 4.12). In this case the traffic calming circle must be fully traversable. The outer radius of the traffic calming circle depends on the dimensions of the largest vehicle (in general the outer radius needs to be about 10 m), and on the diameter of the central island as a function of the outer radius (in general about 4 m). Positive aspects of this type of traffic calming are high speed reduction and recognizable profiles at intersections. This type of traffic calming circle also needs physical and visual support. It has an aesthetic (enhancement of the neighborhood with trees, bushes, flowers), a traffic safety role (by reminding drivers they are in a residential area), and also an informatics role (Fig. 4.13).

Fig. 4.12 Landscaped traffic calming circle; Enna, Italy

Fig. 4.13 Traffic calming circle; Enna, Italy

Traffic calming circles are appropriate solutions at the intersections of local streets or intersections of local streets with collector streets. They can be installed at three-arm (T or Y) and also four-arm (+) intersections, but are most effective (and least expensive) when constructed at a four-arm intersection. If they are installed at three-arm intersections, additional narrowing or widening of roadway is usually necessary in order to achieve the desired results. In several European countries a traffic calming circle is an appropriate solution at intersections with 400–600 PCU/peak hour, and if V_{85} is equal or less than 40 km/h.

4.5 Traffic Signal Controlled Roundabouts

A one-lane roundabout normally performs satisfactorily when entry flows are reasonably balanced and this is usually a consideration regarding intersection choice. However, large roundabouts have sometimes been created as a result of multiple entry arms. Congestions at these roundabouts can be caused by peak traffic conditions, usually when major and minor flows are unreasonably balanced or at high circulating traffic speeds on large roundabouts.

In these circumstances one of the possible solutions could be the installation of traffic signals in order to counteract predictable operating imbalance by creating gaps in the circulating traffic (another solution could be one of different "alternative types of roundabout").

We could also have situations where a tramline or light rail intersects a roundabout (its arm or central island). In such cases, one of the possible solutions could be the installation of traffic signals, and another could be to up-grade traffic management (over- or under-pass).

Traffic signal controlled roundabouts originate from the UK and go back to the early seventies of the past century. The first experiment regarding signals at a roundabout in the UK was in 1959, when traffic signals were initially installed at roundabouts as part time signals operating during peak periods, and this application is still common. Basically, signalized roundabouts were formally recognized in 1984 as a possible alleviation of overloading or unbalanced flow, caused by differential traffic growth. Traffic signals could be installed at some or all of the entry points, to operate either continuously or at peak hours only. A few years later, in 1990, the installation of traffic signals at roundabouts (particularly grade-separated) had become a popular low-cost remedial measure for dealing with excessive queues and congestions during peak periods [6].

However, not until 1991 can we speak of their rapid expansion. From that year onwards, traffic signals became a popular method of traffic control at roundabouts and are now also well-known in Belgium, The Netherlands, Denmark, Sweden, Germany, Poland and other European and also non-European countries (e.g. Mexico, Australia).

There are several conditions within the real world where traffic signals at roundabouts are desirable or even necessary. At larger roundabouts with multiple entry arms, we may have situations [6]:

- when a minor flow to the left of the major flow is dominant on the circulatory carriageway;
- when the major flow dominates the circulatory carriageway to the extent that the remaining arms of the roundabout experience severe difficulties;
- when a large U-turn flow severely reduces access from other approaches;
- high circulating traffic speeds at large roundabouts, which may make it difficult for other traffic to enter.

Under these circumstances traffic signals can be installed at roundabouts to counteract predictable operating imbalance by creating gaps in the circulating traffic flow. They can also be used to prevent queue lengths causing problems at adjoining junctions or blocking a motorway from slip-lanes [6].

Nowadays, in several countries it is appropriate or even necessary to implement traffic signals:

- at roundabouts where traffic conditions changed after their implementation;
- at existing, traffic overloaded roundabouts;
- where a tramline or a light rail intersects a roundabout;
- to increase traffic safety regarding heavy volumes of pedestrians and cyclists.

Traffic signals are one of the possible solutions when traffic conditions have changed after the roundabout's implementation. Most often in this situation a predominant traffic flow is created, which means that one of the main criteria for implementing a roundabout does not exist anymore. In these cases, traffic signals are usually implemented, whilst the other two options are the implementation of slip-lanes or up-grading the main traffic direction.

At existing traffic-overloaded roundabouts, traffic signals can be implemented in different ways, depending on the position of a traffic-light installation, the time of traffic-light operation, and the number of signalized arms.

In regard to places for traffic signals' installations, we need to know which vehicles are controlled by traffic signals. Thereby, we have options between direct and indirect control. Under direct control, traffic signals control the vehicles entering and exiting the roundabout. Therefore, the traffic signals are located at the roundabouts' entries and along the circulatory carriageway. Under indirect control, traffic signals control only the vehicles at the roundabouts' entries and are not located along the circulatory carriageway.

Traffic signals at a roundabout can operate continuously or part-time only. It is also possible to implement a ramp metering system [7], which means that vehicles are released into the circular flow at intervals, controlled by a ramp meter, and connected to detector loops. In this case, traffic signals' devices operate part time only. A ramp metering system serves as a traffic regulation measure. It was first used on highways and express roads and is well-established in the USA, Australia, New Zealand, the UK, Germany, France and The Netherlands. Its purpose is to enable easier and safe merging of vehicles onto highways, express roads or other sections burdened with traffic. A ramp meter connected to detector loops controls the releases of vehicles at time intervals. A ramp meter placed alongside or over the carriageway is connected to an induction loop serving as a detector and is installed underneath the carriageway, is traffic dependent and only operates during peak traffic or during traffic congestion periods.

The number of traffic-signalized arms is also a very important criterion concerning the type of traffic signalization. Firstly we need to know how many arms need to be signalized (because it is unnecessary for all arms to be signalized). In some cases only one arm of a roundabout is signalized.

If a tramline or light rail intersects a central island or one arm of a roundabout (Fig. 4.14), traffic signals are one of possible solutions for preventing conflicts between the road and rail transport (another solution could be to up-grade traffic management—over- or under-pass). Thereby, we need to ensure that vehicles driving through a roundabout do not stop on a railway line. Such solutions are quite common and can be found in The Netherlands, Belgium, France, Sweden, Norway, Poland, Italy and in several other countries.

Traffic signals can be implemented in different ways. The differences are mostly in the placements of light-signaling devices, and on their types (Fig. 4.15).

Light-signaling devices can be mounted at a roundabout's entries or along the circulatory carriageway itself, before the crossing points of road and rail transport.

Different signaling devices can be used, either traffic lights with three-colored light signals or a light-signal device when only a red light turns on when the tram or railway approaches. When road vehicles have a free path, the light-signal device is off [8].

When a tramline or a light rail intersects a roundabout's arm, the following two methods are applied:

- stopping all vehicles at a roundabout's entries (Fig. 4.16); and
- stopping vehicles only at the road and railway crossing (Fig. 4.17).

A high volume of pedestrians and cyclists diminishes a roundabout's capacity. It may cause problems by filling-up and locking a roundabout, and may also influence the traffic safety level. We can increase traffic safety regarding high volumes of

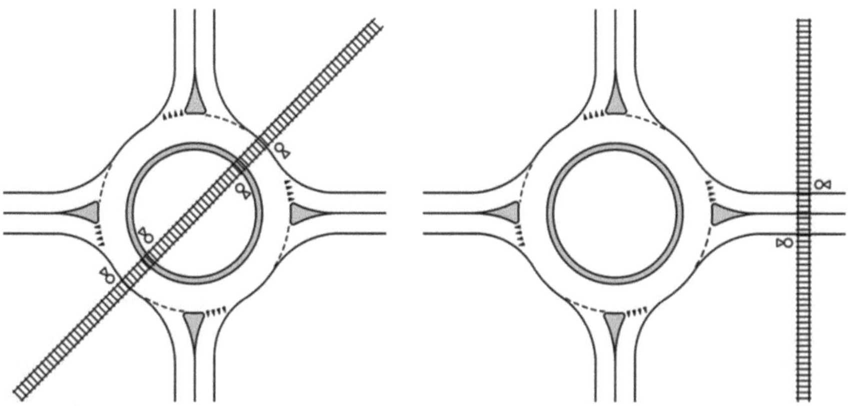

Fig. 4.14 Tram or suburban railway intersects a central island and an arm of a roundabout

Fig. 4.15 Traffic signals at a roundabout with a light rail crossing; Rotterdam, The Netherlands

Fig. 4.16 Stopping all vehicles at a roundabouts' entries

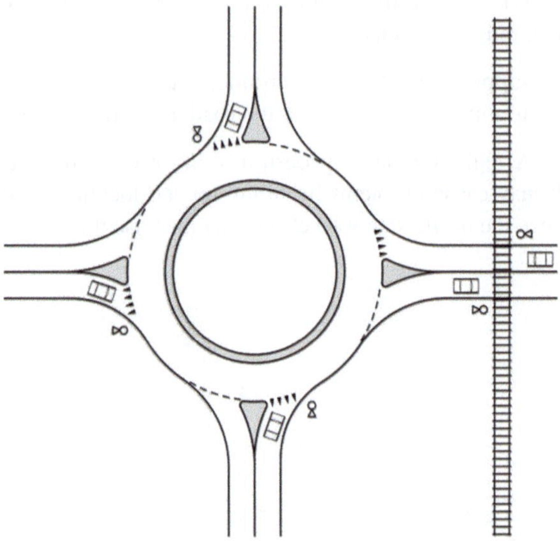

pedestrians and cyclists (especially at large roundabouts) in two ways, by up-grading the management of cyclists and pedestrians, and by implementing traffic signals.

Pedestrian-crossing signalization at roundabout can be implemented in different ways. Where traffic lights regulate only the flow of pedestrians and cyclists, it is sensible to implement traffic lights with a push-button, which guarantee safe passage

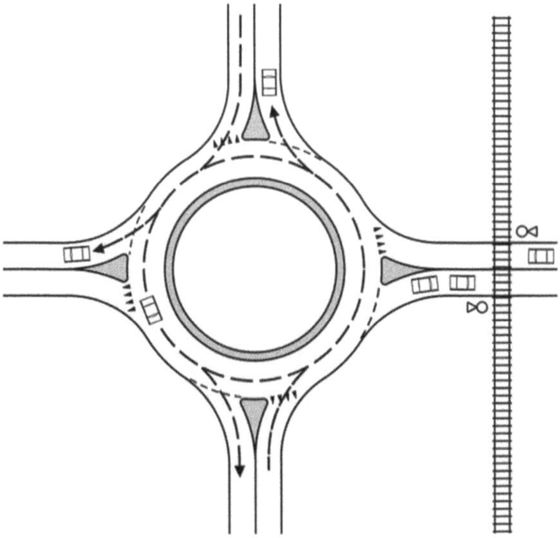

Fig. 4.17 Stopping vehicles only at the road and railway crossing

across the roundabout's arm. Normally, pedestrian and cyclist crossings run together, and in those cases the same traffic light may regulate both the pedestrian and cyclist traffic. But, there also exist other more complicated solutions and it seems that the best experience with traffic signals on roundabouts exists in France [8].

As should be clear from the above, traffic signals at roundabouts is one of the complex topics, with a lot of possibilities for development in the future.

4.6 Turbo-Roundabout

4.6.1 Introduction

There is no doubt that Dr. ir. L.G.H. Fortuijn, now a researcher of the University of Delft, The Netherlands, is the "father" of the turbo-roundabout (Fig. 4.18). And there is also no doubt that this type of roundabout has been a more popular alternative or unconventional type of roundabouts over the last decade all over Europe. It seems that a turbo-roundabout is a kind of "fashion" nowadays.

The turbo-roundabout was developed in 1996; the first turbo-roundabouts were installed at the end of the 1990s in The Netherlands, by the end of 2007 there were seventy turbo roundabouts [9], whilst at the end of 2013 there were more than 200 turbo-roundabouts in The Netherlands [10].

The turbo-roundabout was primarily developed to deal with entering and exiting conflicts that occur on standard two-lane roundabouts of the type that is frequently in use in several European countries. Since the introduction of the turbo roundabout, standard two-lane roundabouts are no longer being constructed in

Fig. 4.18 The Netherlands' typical turbo roundabout

The Netherlands. The idea of the turbo-roundabout was very rapidly (just over a few years) transposed into several countries such as Slovenia [11], Germany [12], Denmark, and Czech Republic [13], as also Hungary, the Former Yugoslav Republic of Macedonia and several other countries.

Experiences provide good insight into the effects on road safety, capacity, and experience by road users in some countries but also less satisfactory experiences in other countries because of the already known reason: Road marking (without divided curbs) does not prevent the change of traffic lanes at the turbo-roundabout!

4.6.1.1 Short Explanation of a Turbo-Roundabout

A turbo-roundabout (Fig. 4.19) is a special type of (usually) two-lane roundabout, where some directed traffic flows are separated or run along physically-separated lanes [14], with multiple centers of outer and inner diameters and traffic lanes (spiral course of a carriageway). Traffic flows run separately at turbo-roundabout even in front of actual entries onto a roundabout, run separate lanes throughout the roundabout, and when exiting the roundabout lanes are separated again. The physical separation of traffic lanes is interrupted only at places of entry onto the inner circulatory carriageway.

The most important element of a turbo-roundabout is the divided curb (delineator) for eliminating the necessity of weaving. This results in both an increase in road safety as well as in the capacity of a roundabout (Fig. 4.20).

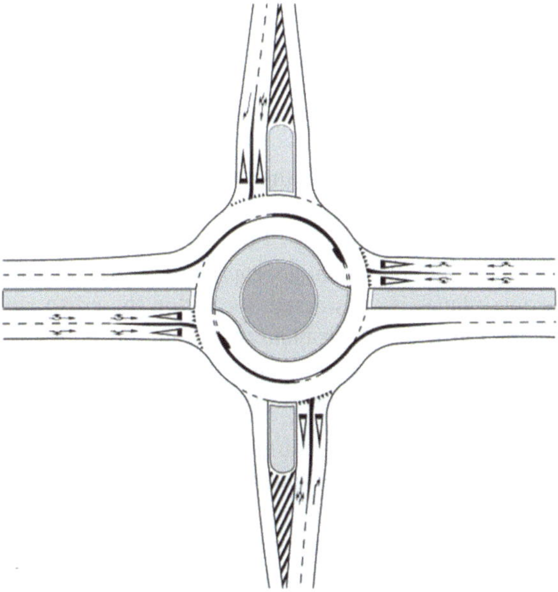

Fig. 4.19 Typical layout of a turbo-roundabout [14]

Fig. 4.20 Divided curbs (delineators) on a turbo-roundabout's circulatory carriageway

As a result of lane dividers, drivers need to choose a correct lane before they enter a roundabout; therefore drivers should be assisted by clear signposting and lane markings. A special form of signposts and arrow-markings has been developed for this type of roundabout for easier and clearer lane selection in front of roundabout (Fig. 4.21).

Some experiences [11] have proved that a turbo-roundabout is an appropriate solution at locations outside the urban areas, normally using one main and one side traffic route, regarding the intensity of the traffic flow. A turbo-roundabout in an urban area is only conditionally an appropriate solution. If a roundabout with two entry and two exit lanes is considered, located in the urban areas, we must at first solve the leading traffic-safety problem of non-motorized participants.

Experiences differ from country to country but, in general, a turbo-roundabout is conditionally an appropriate solution in the cases of:

- existing traffic-overloaded one-lane roundabout, the size of which (outer radius) enables the implementation of an additional circulatory lane inwards (better solution) or with enough space for the implementation of another circulatory lane outwards (somewhat less appealing and more expensive solution);
- existing traffic-overloaded two-lane roundabout;
- existing traffic less safe two-lane roundabouts;
- reconstruction of a classic intersection with a predominant main traffic direction and with a heavy traffic flow.

In all these cases, the selection of the turbo-roundabout type also depends on the predominant direction of the main traffic flow. Namely, the predominant

Fig. 4.21 Special form of signposts and arrow-markings at a turbo-roundabout; Slovenia

direction of the main traffic flow is a criterion for the selection of the turbo-roundabout type. Consequently, different types of turbo-roundabout were developed for specific combinations of traffic volumes and directions [9].

A geometrical form of the turbo-roundabout is a little bit complicated as it is formed by the so-called turbo block (Fig. 4.22). This is a formation of all the necessary radii, which must be rotated in a certain way, thereby obtaining traffic lanes or driving lines.

The center of a turbo block must be located in such a way that a radial connection of all entries into the roundabout with a spiral course of a circulatory carriageway is possible. The turbo block also contains (besides all radii) the so-called translator axle. A translator axle is an axle, where a shift (movement) of different radii occurs. A shift of radii depends on the width of the circulatory traffic lane and on the locations of the verges [9].

The best position of the translator axle is as if the clock hands pointed to "five minutes to five o'clock" (Fig. 4.23) in the case of a four-arm or "ten minutes past eight o'clock" in the case of a three-arm turbo-roundabout [9].

The size of radii of a turbo-roundabout and the width of the circulatory carriageway must be selected in such a way that the driving speed through the roundabout does not exceed 40 km/h (Table 4.1).

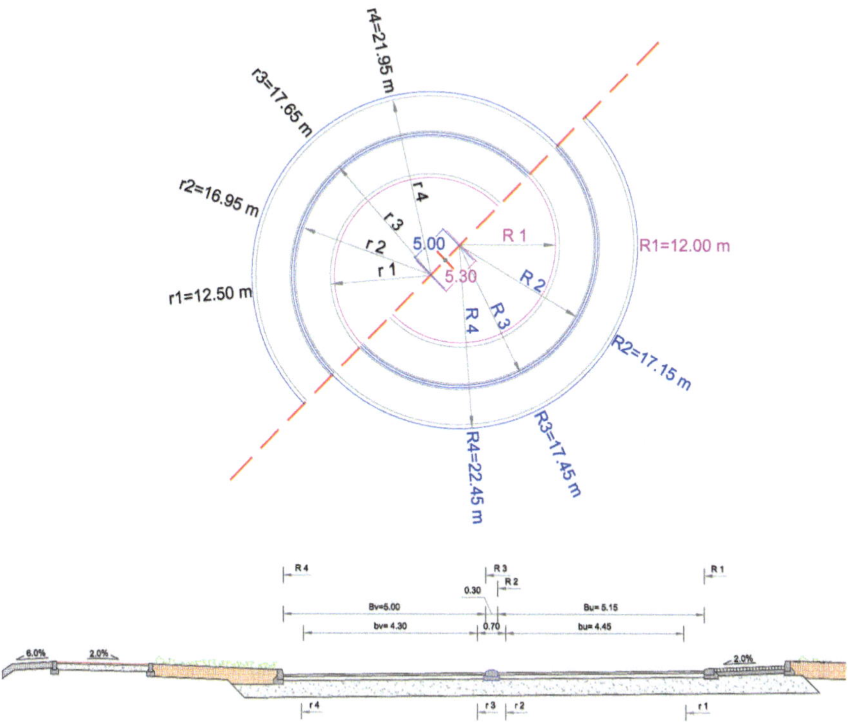

Fig. 4.22 A turbo block with a translator axle

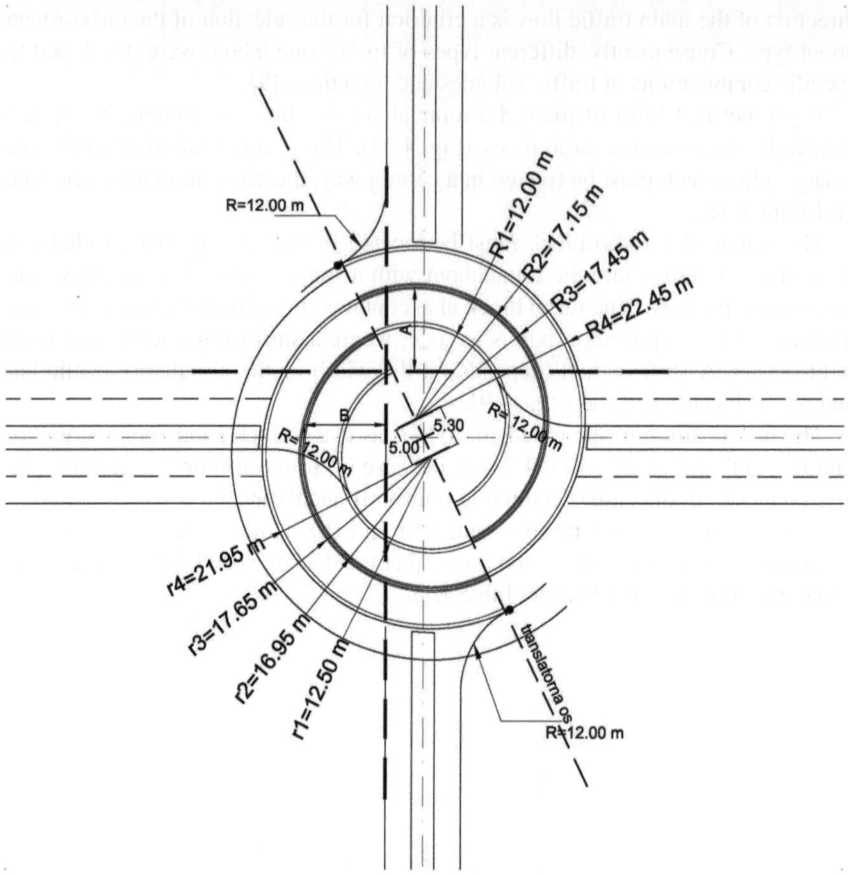

Fig. 4.23 The best position of the translator axle in the four-arm turbo-roundabout (normal size)

A turbo-roundabout has a higher level of traffic safety in comparison to a "standard" two-lane roundabout for several reasons. The most important is a lower number of conflict spots (Fig. 4.24). Entering and exiting conflicts are eliminated by directing motorists to correct lanes before entering a roundabout and introducing spiral lines that guide motorists to the correct exit. A turbo-roundabout reduce the number of conflict spots of crossings (by reducing the number of crossing traffic flows), and eliminate weaving conflict spots (by the separate running of individual direction flows). A further benefit is that traffic in the main direction has to consider one lane only before entering a roundabout.

Reduction of a number of conflict spots at turbo-roundabout, compared with a "standard" two-lane roundabout, causes a global reduction of crash probabilities. One should note that some of these conflicts have a higher severity, not only because of the increased impact angle, but also because circulating traffic is sometimes concentrated at the outer lane. In the absence of crash data history, this research field justifies in depth analysis, using micro simulation techniques.

Table 4.1 The size of radii of a turbo-roundabout [9]

Size of the turbo roundabouts elements (m)				
Element	Mini	Normal	Medium	Large
R_1	10.45	12.00	14.95	19.95 (21.70)
R_2	15.85	17.15	20.00	24.90 (27.10)
R_3	16.15	17.45	20.30	25.20 (27.40)
R_4	21.20	22.45	25.25	29.95 (32.80)
r_1	10.95	12.50	15.45	20.45
r_2	15.65	16.95	19.80	24.70
r_3	16.35	17.65	20.50	25.40
r_4	20.70	21.95	24.75	29.45
B_v	5.05	5.00	4.95	4.75 (5.40)
B_u	5.40	5.15	5.05	4.95 (5.40)
b_v	4.35	4.30	4.25	4.05
b_u	4.70	4.45	4.35	4.25
D_v	5.75	5.30	5.15	5.15 (5.50)
D_u	5.05	5.00	4.95	4.75 (5.50)

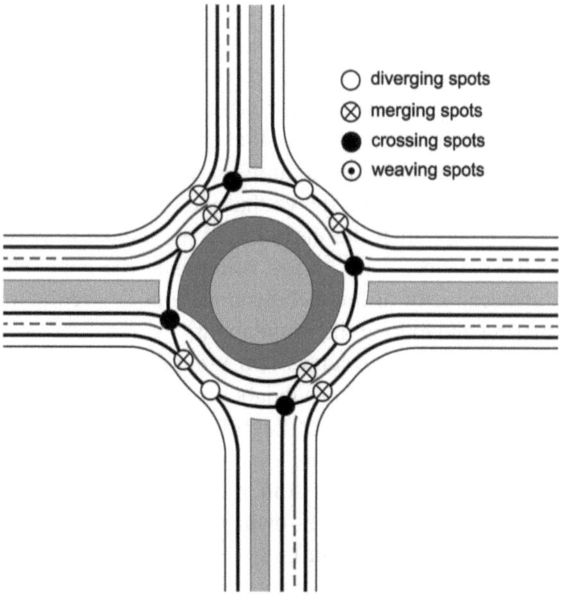

Fig. 4.24 Conflict spots at a turbo-roundabout with two lane entrances and exits on major roads, and two lane entrances and one lane exits on minor roads

Some studies have shown 70 % lower crash risk if a "standard" two-lane roundabout is reconstructed as a turbo-roundabout [14]. Other studies, based on conflict analysis techniques applied to 9 layouts with different demand scenarios, showed 40–50 % reductions in accident rates [15]. In the study based on micro simulation applications, Fortuijn [16] concluded that drivers using the outer lane of a turbo-roundabout drive slower than in the "standard" two-lane roundabout, with reductions from 48 to 38 km/h.

Importantly, research from The Netherlands makes a comparison between turbo-roundabouts and traffic signal controlled or yield intersections. It shows that a 70 % reduction in accidents resulting from serious injuries can be expected when introducing a turbo-roundabout at such an intersection [14]. The same applies to the introduction of a one-lane roundabout, but obviously this would result in a lower roundabout capacity.

A turbo-roundabout has a larger capacity compared to a "standard" two-lane roundabout for several reasons:

- the roundabout entries of this type are usually two traffic lanes, which directly continue into two circulatory traffic lanes;
- use of the inner circulatory traffic lane becomes more attractive, since there is no need for weaving;
- the entry traffic flow is no longer hesitant when entering the circulatory carriageway, which increases the capacities of entries.

Although the safety benefits are widely recognized, there are still some doubts about improved capacity in some countries. The main reason is that practical evaluation data is presently unavailable for turbo-roundabouts because only in The Netherlands have a number of turbo-roundabouts been realized and very few of those are operating at (or near) capacity. But, without doubt, turbo-roundabout offer better capacity than "standard" roundabout of similar size. The quick-scan model for capacity evaluation, as developed by the Province of South Holland, The Netherlands [9], shows that the capacity of a turbo-roundabout is from 25 to 35 % higher than the capacity of a "standard" two-lane roundabout, depending on the balance of the traffic volumes at the approaches. The main reason for the higher capacity of the turbo-roundabout is a reduction of conflict spots for traffic entering and exiting the roundabout.

It seems that nowadays two groups of countries exist with turbo-roundabouts: countries having turbo-roundabouts with divided curbs (The Netherlands, Slovenia, Hungary, and the Former Yugoslav Republic of Macedonia), and without divided curbs, using road markings only (Germany, Denmark, Czech Republic, Lithuania, Canada, and the USA). And it seems at the time of writing this book that just two countries have their own guidelines for designing turbo-roundabouts (The Netherlands and Slovenia).

As The Netherlands' experiences with turbo-roundabouts are already "deja-vu", the following text describes briefly some of the results published on the turbo-roundabouts as experienced in some other countries. Namely, during the last decade several turbo-roundabouts have been constructed in several countries, with both good and less-well results, depending on design elements of their turbo-roundabouts.

4.6.2 Slovenian Experiences

The idea of a turbo-roundabout was very rapidly (in a few years) transposed into the Slovenian environment for several reasons. One of the more important reasons was surely the fact that in the past undersized two-lane roundabouts had been constructed in Slovenia. The second of the more important reasons was that inner circulatory lanes were unacceptable by younger and senior drivers because they felt insecure when changing lanes on a circulatory carriageway [11]. Accordingly, the first ideas on the implementation of turbo-roundabouts began to appear in Slovenia in 2002 (Fig. 4.25).

The first Slovenian-turbo roundabout was installed in the city of Koper (Fig. 4.26), which is often called "city of roundabouts", as there are just two traffic signal controlled intersections in the entire city.

At the time of writing this book (April 2014) there are eleven existing turbo-roundabouts in Slovenia, two assembled turbo-roundabouts (Fig. 4.27), one traffic-lighted turbo-roundabout (Fig. 4.28), two turbo-roundabouts are under construction, and design documentation for four more turbo-roundabouts being processed.

It needs to be stressed that a Slovenian typical turbo-roundabout differs from The Netherlands' typical turbo roundabout. Certain dimensions have been changed in order to meet Slovenian conditions. A typical Slovenian turbo-roundabout is smaller, they use a different design element for weaving prevention (usually without "peaks") because of snow plowing (Fig. 4.29), in urban areas intermediate

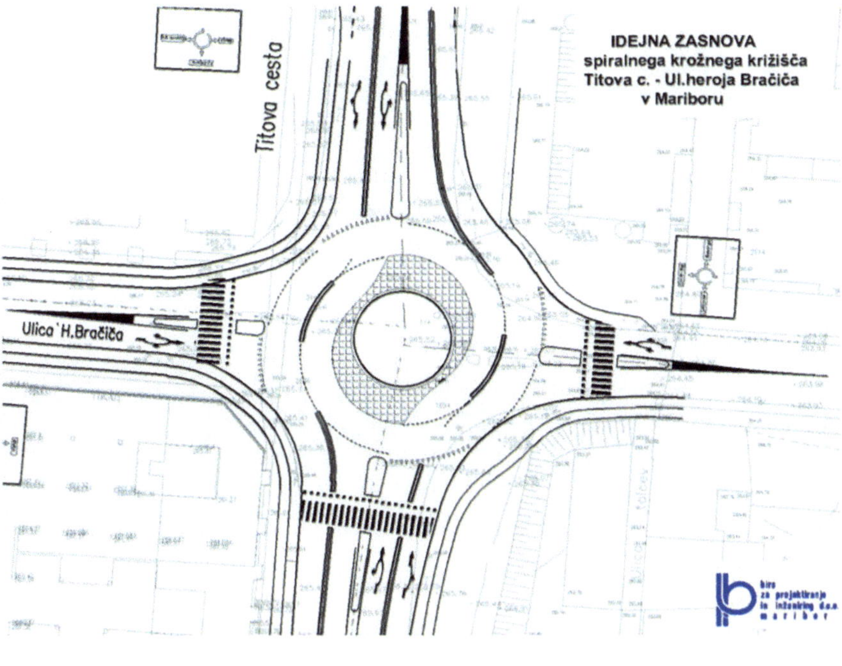

Fig. 4.25 The first Slovenian idea that included a turbo-roundabout

Fig. 4.26 The first Slovenian turbo roundabout, the city of Koper

Fig. 4.27 Slovenian assembled turbo-roundabout; Koper

Fig. 4.28 Traffic-lighted turbo-roundabout; Ljubljana

Fig. 4.29 Divided element (delineator) without a "peak" on Slovenian turbo-roundabout

splitter islands between two lanes at entrances and exits (Fig. 4.30), and a different design for entrances to inner circulatory lanes (see Fig. 4.29).

An analysis of traffic safety on Slovenian turbo-roundabouts was carried out at the end of 2013, only for those turbo-roundabouts that were built through the reconstruction of previous (or existing) intersections, since this was at that moment the only way to compare the situations before and after reconstruction. From the available data on traffic accidents, it could be seen that at some intersections serious traffic accidents had occurred before their reconstruction, and that no such accidents had occurred since the construction of turbo-roundabouts. Consequently, it could also be concluded that the reconstructions of non-signalized and signalized intersections into turbo-roundabouts in Slovenia had been justified—at least from the traffic safety point of view.

In general, turbo-roundabouts in Slovenia have met the expectations concerning the large capacity and particularly the high level of traffic safety. In drivers' opinions, this type of roundabout is very safe due to the following reasons:

- driver is on his "own" lane all the time;
- there is no weaving on circulatory carriageways;
- it is always clear who has the priority;
- no fears and doubts when driving on inner circulatory carriageways;
- lower speed compared to "common" multi-lane roundabouts;
- signposts and road markings are easily understood and unmistakable.

The same good experiences with turbo-roundabouts are their capacities, no bottlenecks, with daily traffic between 38,000 and 42,000 vehicles per day.

Fig. 4.30 Intermediate splitter lane between two driving lanes on entrance and exit on Slovenian turbo-roundabout in urban area

4.6.3 Czech Experiences

It is unknown exactly when the idea of designing turbo-roundabouts in the Czech Republic was introduced [17]. Probably the first turbo-roundabout ever built in the Czech Republic was the one in Modřice near Brno in 2006, and shortly after another one in Brno next to Masaryk university campus, in 2007 (Fig. 4.31). Currently 10 turbo-roundabouts have been built nationwide and approximately 20 are under construction.

There are two major reasons for building turbo-roundabouts in the Czech Republic.

The first one is traffic safety. It is now a well-known fact that two-lane roundabouts have caused a large number of traffic accidents. The causes of accidents on these two-lane roundabouts in the Czech Republic are similar as in other European countries. It is especially caused by the unexpected change of lanes, both on the circulating carriageway and at the entry or exit, and intense braking. The problem is in several conflict spots and substantially different speeds of vehicles driving on the circulating carriageway, compared to those at the entries to the roundabout. The second reason is capacity. After reaching the capacity of one-lane roundabouts, two-lane roundabouts were designed with the goal of increasing this capacity. However, these expectations were not met. In reality, although there was indeed an overall increase in capacity between a one-lane and two-lane roundabout with a standard layout, it tended to be only about 10 %. The inside lane was used very rarely or the drivers used it only on the actual circulating carriageway when passing other vehicles which exit before them [18].

In general, there were three problems they had to deal with when designing turbo-roundabouts in the Czech Republic. The first one was road marking, dividing or

Fig. 4.31 Turbo-roundabout in Brno; Czech Republic

rather not dividing the individual driving lanes with concrete divided curbs, the width of the lane at the entry, circulating carriageway and exit of the turbo-roundabout, which are associated with dividing or not dividing elements [19].

Currently, changes in road markings in their technical standards TP169 Rules for road signing [19] are at the moment of writing this book just about to be approved. The road signing will be similar to the Slovenian.

The second problem relating to the design and construction of turbo-roundabouts in the Czech Republic was related to the physical divisions of individual driving lanes upstream, inside and downstream of the roundabout (Fig. 4.32), similarly to that done in The Netherlands and Slovenia. The main reason why the relevant governmental authorities do not allow physical division of driving lanes inside turbo-roundabouts, is allegedly the fact that it would not be possible to maintain during the winter and that concrete delineators could be a dangerous element for two-wheeled vehicles [19].

The third problem was the widths of lanes at the entries to the circulating carriageways of the roundabout (see Fig. 4.33), at the circulating carriageway itself and at the exit from the roundabout. It can be seen that the widths of the driving lanes are insufficient for trucks, which thus have to partially drive into the neighboring lane. Also the radii of the curves at the entry and exit from the circulating carriageway are too small. This proposal does not accept sufficient extensions of driving lanes at two-lane entries onto the roundabout. If such an extension is based on the trailing lines of the largest vehicle, in the absence of their physical division the roundabout experiences quick direct bypasses of vehicles (Fig. 4.34).

Fig. 4.32 Turbo-roundabout, truck passing through; Olomouc, Czech Republic

Fig. 4.33 Turbo-roundabout with right-hand turning bypasses; Olomouc, Czech Republic

It should be emphasized that there is no such Czech typical turbo-roundabout because each example differs from another. Czech turbo-roundabouts are planned and constructed in order to meet the specific conditions and needs at a specific place without any division into some specific, well described groups and their parameters (see very different examples in Figs. 4.33 and 4.34).

It can be said that the currently existing turbo-roundabouts in the Czech Republic are not used to their full potential in terms of their capacities as it is relatively new approach.

On the other hand, we can say that the application of turbo-roundabouts in the Czech Republic is growing steadily and that these very specific elements of road design will thus contribute to improving traffic safety whilst maintaining the same or higher capacities compared with transversal or 3-way intersections [20].

4.6.4 German Experiences

Standard multi-lane roundabouts, especially with two-lane exits, are not recommended in Germany—neither by guidelines nor by experts' advice—due to the experiences that two-lane exits systematically cause a large number of accidents. Due to the interactions of circulating flows with fast vehicles leaving the circle from the inner lane, these two-lane exits are often a reason for accidents. Therefore, multi-lane roundabouts are not recommended for application in Germany. Also

Fig. 4.34 Turbo-roundabout in Hradecká and Žabovřeská streets; Brno, Czech Republic

three-lane (or larger) non-signalized roundabouts are not under consideration in Germany [12]. Because of this, it could be expected that turbo-roundabouts will come up also in Germany.

The first turbo-roundabout in Germany has been opened in 2006 in the town of Baden-Baden (Fig. 4.35). Here, *vis-a-vis* from the major entries, a second lane is added on the inner side of the ring, whereas at exits with a significant exiting flow the vehicles on the outside lane are forced to continue their way onto the exit [12].

It seems that the initial safety data has not been satisfied, mainly because of priority accidents at the high speed entrances. Here, initially the priority movements on the circular lane had only a very small volume with the consequence that the approaching drives had to observe priority quite seldom and thus were not prepared to stop. Meanwhile, after 7 years of operation, this problem has been settled due to increasing traffic on the circle.

The turbo-roundabout in Baden-Baden (like the others in Germany) do not use raised lane-dividers. Thus crossing of the lane marking is possible (see Fig. 4.36). However, this has not turned out as a significant reason for accidents. Therefore, vertical lane-dividers are not required and consequently, they are not recommended in Germany for turbo-roundabouts.

In Germany, as the most important feature for safety it is recommended to avoid pedestrian and bicycle crossings of the turbo-type entrances and exits of turbo-roundabouts. Recently a research project on such turbo-roundabouts was completed. The result was that turbo-roundabouts offers a potential to combine a level of safety like the compact roundabouts but at larger capacities [22].

Fig. 4.35 The first turbo-roundabout in Germany; Baden-Baden [21]

Fig. 4.36 An illegal crossing of lane marking [21]

Meanwhile several examples (e.g. Zweibrücken, Achern, Tuttlingen) of this type of roundabout have been built and are operated successfully.

Fig. 4.37 Typical Danish turbo-roundabout; near Aalborg

Recently a committee by FGSV, the organization which develops guidelines and standards regarding highways and road traffic, has been developed a specific guideline for turbo-roundabouts [23].

4.6.5 Other Countries' Experiences

According to the web page of Dirk de Baan [24], at the present there are 320 turbo-roundabouts all over the world, although it is questionable as what in some countries they understand under the terminus of a turbo-roundabout.

Notwithstanding the foregoing, according to his research there are also turbo-roundabouts in Poland, Denmark (Fig. 4.37), Romania, Hungary (Fig. 4.38), Canada, Austria, as also in South-Africa (Ongoye).

4.7 Dog-Bone Roundabout

The dog-bone roundabout (Fig. 4.39) (due to its aerial resemblance to a toy dog bone), supposedly invented in The Netherlands [9], is a variation of the dumb-bell roundabout. A dog-bone is also a hybrid like a dumb-bell, combining a diamond interchange and a roundabout. Dog-bone occurs when a roundabout does not form

Fig. 4.38 Hungarian turbo-roundabout; near Szeged

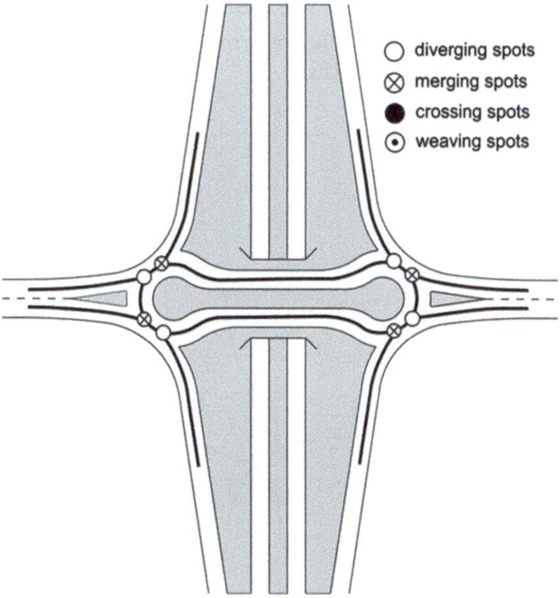

Fig. 4.39 Dog-bone roundabout; sketch

a complete circle but has a "raindrop" or "teardrop" shape instead. These two roundabouts are fused together, forming a single "squashed" roundabout; by merging two central islands an oval one-lane roundabout is obtained. The interior of a roundabout is closed, and parallel roadways are used between ramp terminals.

A dog-bone roundabout is a better solution than a "standard diamond interchange" for several reasons. At a dog-bone roundabout speeds are significantly lower than at a standard diamond interchange, as two roundabouts are a measure for traffic calming. At a standard diamond interchange drivers might make a mistake and turn into the wrong direction at the ramp, while at a dog-bone roundabout such an option is significantly lower. A dog-bone roundabout, like the dumb-bell, even provides the possibility for completely eliminating the option of driving in the wrong direction, using the adequate deflection of a ramp.

A dog-bone roundabout can generally handle traffic with fewer approach lanes than other intersection types. Direct U-turn is impossible, although this maneuver can be made by circulating around both roundabouts.

This type of roundabout is slightly more expensive than a dumb-bell roundabout because of three lanes—instead of two lanes as in the case of the dumb-bell—on a flyover, but still cheaper than a diamond interchange. It could be constructed as two "standard" one-lane roundabouts or as two turbo-roundabouts.

This configuration reduces conflicts between vehicles entering dog-bone roundabouts from ramps, increasing roundabouts' capacities, reducing queuing and delays, when compared with a diamond interchange.

At a dog-bone roundabout there are just 4 merging and 4 diverging conflict spots (as on the one-lane roundabout) (Fig. 4.40).

This type of roundabout has started to be very common in different European countries, especially in The Netherlands (Fig. 4.41) and in the UK, and the idea is very rapidly transposing itself into other European (Fig. 4.42) and non-European countries (also in the USA).

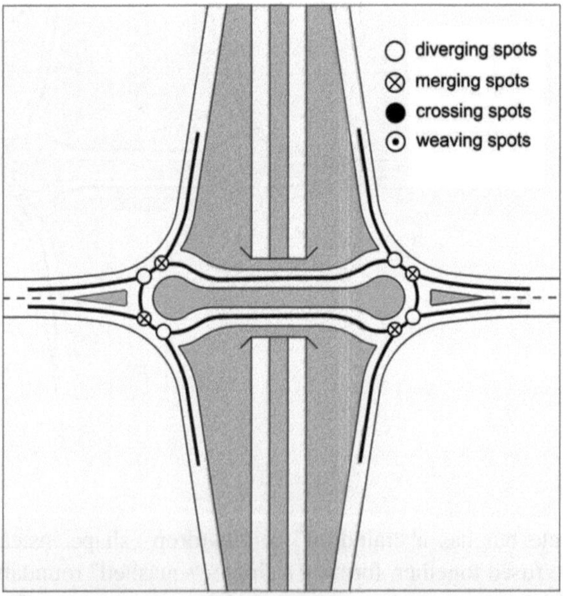

Fig. 4.40 Conflict spots at a dog-bone roundabout

Fig. 4.41 Dog-bone roundabout formed by two turbo-roundabouts; near Den Haag, The Netherlands

Fig. 4.42 The first Slovenian idea for a solution that included the dog-bone roundabout; city of Koper

Examples of dog-bone roundabouts in the USA are located along Keystone Parkway in Carmel (Indiana), Avon (Colorado) and at some other places.

References

1. Tollazzi, T., Renčelj, M. & Turnšek, S. (2011). Slovenian experiences with alternative types of roundabouts-"turbo" and "flower" roundabouts. In *8th international conference environmental engineering*. Vilnius, Lithuania: Vilnius Gediminas Technical University Press Technika.
2. Bezirksregierung Düsseldorf. (2005). *Prioritätenreihungen der Maßnahmen des Landesstraßenbauprogramms* 2005, Düsseldorf.
3. Institute of Transportation Engineers. *Traffic Calming Measures—Neighborhood Traffic Circle*, Washington, USA. Accessed April 2 2014 from http://www.ite.org/traffic/circle.asp.
4. CROW. (1998). *Recommendations for traffic provisions in built-up areas*, Record 15, ASVV, The Netherlands.
5. TSC Krožna križišča. (2011). Slovenian guidelines for roundabout design, Direkcija Republike Slovenijezaceste.
6. Brown, M. (1995). *The design of roundabouts*, TRL, HMSO, UK.
7. Natalizio, E. (2005). Roundabouts with metering signals. Paper presented at the institute of transportation engineers, 2005 Annual Meeting, Melbourne, Australia.
8. Guichet, B. (2005). *Evolution of roundabouts in France and new uses*, Paper presented on national roundabout conference, TRB, Vail, Colorado, USA, May 22–25 2005.
9. CROW. (2008). *Turborotondes*. Publicatie 257, Dutch Information and Technology Platform, The Netherlands.
10. Silva, A. B., Santos, S., & Gaspar, M. (2013). Turbo-roundabout use and design. In *CITTA 6th annual conference on planning research, responsive transports for smart mobility*, Coimbra, Portugal.
11. Tollazzi, T., Renčelj, M., & Turnšek, S. (2011). Slovenian experiences with turbo-roundabouts. In *3rd international conference on roundabouts*, TRB, Carmel, Indiana, USA, May 18–20 2011.
12. Brilon, W. (2011). Studies on roundabouts in Germany: Lessons learned. In *3rd international conference on roundabouts*, TRB, Carmel, Indiana, USA, May 18–20 2011.
13. Súkenník, P., Hofhansl, P., & Smělý, M. (2013). *Turbo-okružní křižovatky: syntéza bezpečnosti a kapacity*, Bezpečná dopravní infrastruktura, Pondělí.
14. Fortuijn, L. G. H. (2009). Turbo roundabouts: Design principles and safety performance. *Transportation Research Record: Journal of the Transportation Research Board, Transportation Research Board, 2096*, 16–24. doi:10.3141/2096-03.
15. Mauro, R., & Cattani, M. (2010). Potential accident rate of turbo-roundabouts. In *4th international symposium on highway geometric design*, Valencia.
16. Fortuijn, L. G. H. (2007). *Turbo-Kreisverkehre—Entwicklungenund Erfahrungen*, Seminar Aktuelle Theme der Strassenplanung, Bergisch Gladbach, Germany.
17. Smely, M., Radimsky, M., & Apeltauer, T. (2011). The capacity of roundabouts with multilane entrances. *Traffic Engineering, 6*(1), 20–21.
18. Smely, M. (2011). *Multilane roundabouts. Civil Engineering*, Vol. 1, No. 6.
19. Smely, M., & Radimsky, M. (2011). Transit of vehicles through multilane roundabout. In *11th international scientific conference MOBILITA* 11, (pp. 278–283), Bratislava, Slovakia, May 26–27 2011.
20. Smely, M., Radimsky, M., & Apeltauer, T. (2011). Traffic safety of multi-lanes roundabouts. In *International scientific conference modern safety technologies in transportation MOSATT* 2011, (pp. 259–264), Kosice, Zlataldka, Slovakia, September 20–22 2011.

21. Brilon, W. (2008). Turbo-Roundabout—an experience from Germany. In *National roundabout conference*, TRB, Kansas City, Missouri, USA, May 18–21 2008.
22. Brilon, W. (2014). Roundabouts: A state of the art in Germany. In *4th International conference on roundabouts*, TRB, Seattle, Washington, USA, 16–18 April 2014.
23. Forschungsgesellschaft für Straßen- und Verkehrswesen. (2014). *Hinweise zu Turbo-Kreisverkehren* (Notes on turbo-roundabouts), Draft version.
24. de Baan, D. (2014). Verkeer—Verkeersveiligheid—Vorm, *Turborotondes—locaties*. Accesed April 7 2014 from http://www.dirkdebaan.nl/locaties.html.

Chapter 5
Alternative Types of Roundabouts at Development Phases

5.1 Introduction

As pointed-out previously, today after many years of experience regarding roundabouts, there are different ideas about the "ideal roundabout". Therefore the development of design rules and advice from an extensive body of research should allow civil and traffic engineers to produce more effective forms of this junction type. It also needs to be stressed that the roundabout has been "at the development phase" since 1902, this development is still in progress, and one of the results of this progress is that several types of roundabouts are in worldwide usage today, called the "alternative types of roundabouts". Some of them are already in frequent use all over the world, some of them are recent and have only been implemented within certain countries, and some of them are still at development phases. It is because of that we can call them "theoretical roundabouts".

The purpose of this chapter is to present just some of them because although we are living within a "global world" this type of information usually needs a couple of years to become known to a wider population. So, there exists a significant probability that at the moment there are more unknown "theoretical roundabouts".

5.2 Roundabout with "Depressed" Lanes for Right-Hand Turning—the "Flower Roundabout"

The roundabout with "depressed" lanes for right-hand turning, in short "the flower roundabout" (see Fig. 5.1), was invented as a solution for achieving a higher level of traffic safety on existing, less-safe standard two-lane roundabouts [1].

One of the basic characteristics of this type of two-lane roundabout is the same as on the turbo-roundabout—physically-separated lanes on a circulatory

T. Tollazzi, *Alternative Types of Roundabouts*, Springer Tracts on Transportation and Traffic 6, DOI 10.1007/978-3-319-09084-9_5

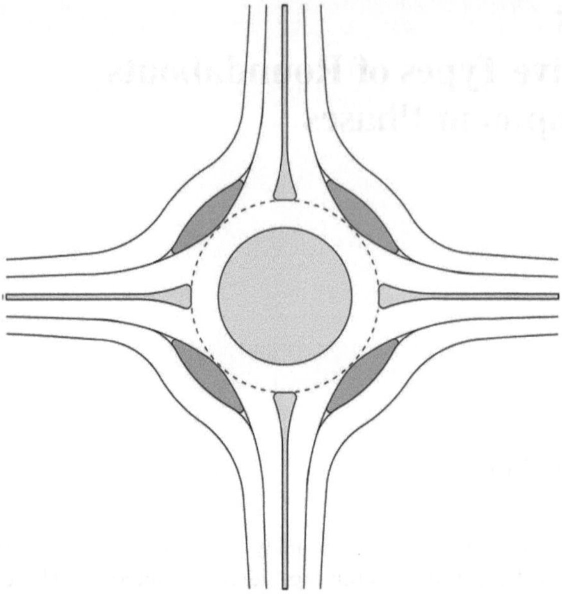

Fig. 5.1 The roundabout with "depressed" lanes for right-hand turners—the "flower roundabout"; sketch

Fig. 5.2 3D rendering of a flower roundabout

carriageway (see Fig. 5.2). The second characteristic of the flower roundabout is that all the right-hand turners have their own separated bypasses (slip lanes). Bypasses caused that the inner circulatory carriageway is only used by vehicles

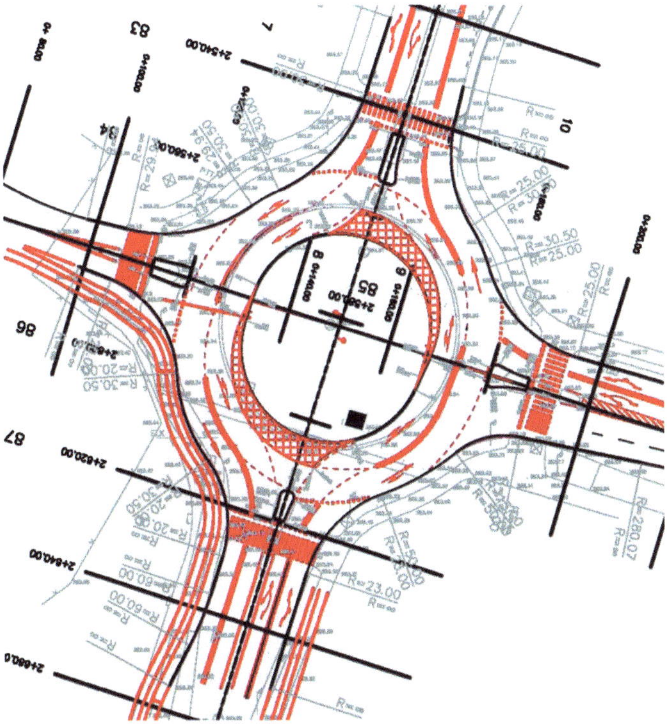

Fig. 5.3 Reconstruction of the existing standard two-lane roundabout into the turbo-roundabout

that are driving through the roundabout (180°), turn for three quarters of a circle (270°) or turn semicircle (360°).

Therefore, bypasses (slip-lanes) for right-hand turners are not a novelty; as they are in frequent use all over the world. A novelty is that it is possible to adjust the existing standard (less-safe) two-lane roundabout into a (safer) flower roundabout without moving any of the outer roads' curbs (see Fig. 5.3), unlike in the case of the turbo-roundabout.

This solution is possible and appropriate on four-lane as well as on two-lane roads. In the case of a two-lane road, an additional, sufficiently long lane is implemented directly in front of an entry/exit.

By physically separating the right-hand turning traffic flow a one-lane roundabout is obtained, with no crossing conflict spots (unlike in the case of the turbo-roundabout), and also no weaving conflict spots (unlike in the case of the standard two-lane roundabout). Any possible weaving conflict spots when transferring from the circulatory carriageway (along the curve) onto the road section (usually as a straight line) are in front of a roundabout (as in the case of the turbo-roundabout), thus being a safer solution from the traffic-safety point of view. In short: at a flower roundabout there are just 4 merging and 4 diverging conflict spots (Fig. 5.4), unlike in the case of a turbo-roundabout with 14 conflict spots [2].

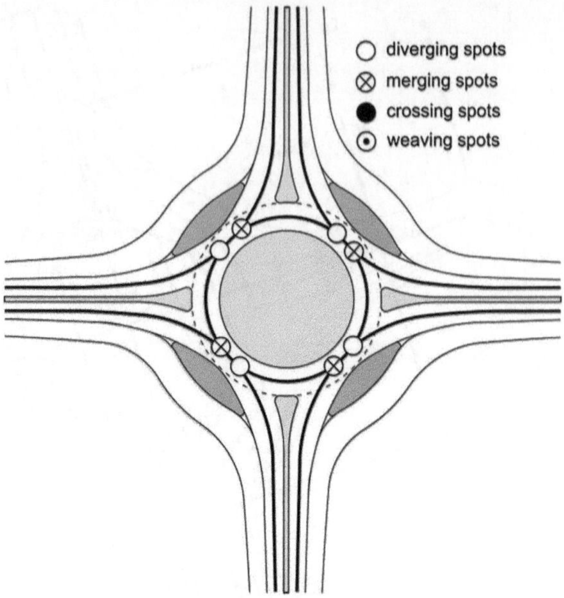

Fig. 5.4 Conflict spots on a flower roundabout

When compared with the standard two-lane roundabout, the main benefits of a flower roundabout are:

- lower number of potential conflict spots between vehicles;
- slower speed along the ring and also along bypasses because of three radii (and not just one);
- elimination of the risk of side-by-side accidents.

In the light of these considerations, flower roundabouts are an alternative to standard two-lane roundabouts, especially for guaranteeing higher safety levels.

Probably the best characteristic of a flower roundabout is that it is implemented within an existing standard two-lane roundabout. Unlike the turbo-roundabout, there is no need to move the outer curbs of the circulatory carriageway; therefore no additional surrounding land is required. When reconstructing a standard two-lane roundabout into a flower roundabout, all the curbs of the circulatory carriageway, splitter islands, and access roads remain in the same positions.

A reconstruction of an existing standard two-lane roundabout into a flower roundabout is performed by four steps (see Fig. 5.5):

Step 1 additional circulatory carriageway towards the center of the roundabout is implemented;
Step 2 construction lines of entries and exits are prolonged;
Step 3 splitter islands are prolonged for one circulatory traffic lane towards the center of the roundabout;
Step 4 redundant surfaces are rearranged into green areas.

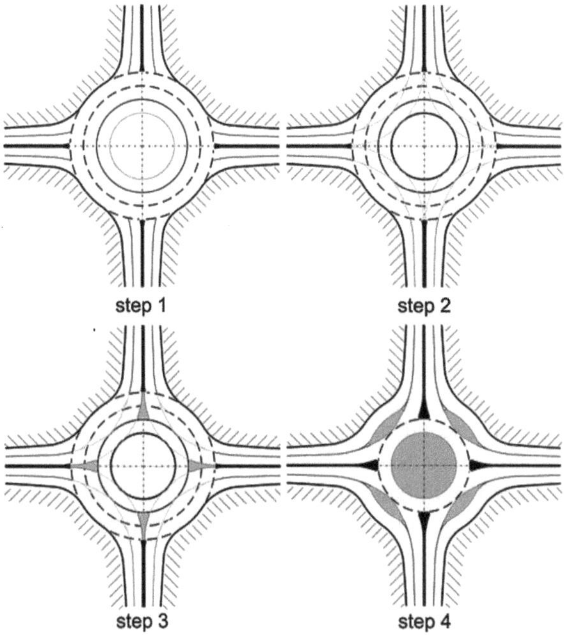

Fig. 5.5 Procedure for reconstructing an existing standard two-lane roundabout into a flower roundabout

It seems that this type of roundabout is very interesting for a variety of research, regarding both capacity [3–7] and environmental impacts [8].

At a flower roundabout the entry capacity (C_E) is a function of the circulating flow in front of the entry under examination (Q_c), the flow exiting from the next arm after that under analysis (Q_u), and the saturation degrees of lanes [6]. Micro simulation software allows the highlighting as to how flower roundabouts, compared to traditional ones with two entry and ring lanes, can be advantageous in terms of capacity (and consequently vehicle delays) when most of the traffic entering the intersection turns towards the right. The results from micro simulation [1] made using VISSIM show that there are no significant differences between turbo-, standard two-lane, and flower roundabouts for low traffic loads. Congestions and queue lengths are approximately the same. It is at higher traffic loads that the differences in favor of the flower roundabout occur, when a percentage of right-hand turners approach 60 % of the total number of vehicles along the main traffic direction [4]. The results of the capacity analyses carried out in [4, 5] have shown that flower roundabouts lead to a significant reduction in delays within all the flow conditions when compared to conventional roundabouts with one entry lane [configuration (1 + 2) or (1 + 1)]. In regard to multi-lane roundabouts (2 lanes at the ring + 2 lanes at entries) flower roundabouts cause higher delays of up to 70 % of the total right-hand turning flows. Once such a threshold is exceeded, flower roundabouts prove to be more convenient than the other schemes (at equal traffic conditions); average vehicle delays decrease more and more markedly.

Guerrieri [6] found that the flower roundabout can be applied to two clearly distinct fields as outlined below:

- modest or moderate circulating flows ($Q_c < 1{,}600$ veh/h): in some combinations of the distribution and intensity of entry flows, a flower roundabout can lead to higher entry capacities than conventional compact or larger-sized roundabouts (2 ring lanes and 2 lanes at entries);
- high circulating flows ($Q_c > 1{,}600$ veh/h): a flower roundabout cannot be used; instead, a conventional large-sized roundabout (i.e. 2-ring lanes and 2-entry lanes) is more appropriate, where circulatory carriageway capacity can reach 2,500 veh/h.

In summary, from the capacity point of view, a flower roundabout can be used whenever the circulating flow is below 1,600 veh/h. If such a circulating flow-value is exceeded, the ring tends to saturate, vehicles can't get onto it, and consequently entry flows reduce towards zero [6].

Road pollutant emissions, above all within the urban context, are correlated to several infrastructural parameters and to traffic intensity and typology. The research work on road junction geometry carried out in European research centers has recently spawned the designing of new road intersection types which are of undoubted interest, especially in terms of traffic functionality and safety [8]. A comparative analysis between the conventional roundabout and the flower roundabout has been carried out in terms of CO, CO_2, CH_4, NO, $PM_{2,5}$ and PM_{10} vehicular emissions, evaluated by means of COPERT Software which has been developed as a European tool for the calculation of emissions from the road transport sector [8]. This model takes into account several traffic and vehicular parameters as: vehicle types, categories and population, annual mileage (km/year), mean fleet mileage (km), etc. The COPERT methodology allows for the calculating of exhaust emissions regarding carbon monoxide (CO), nitrogen oxides (NO_X), non-methane volatile organic compounds (NMVOC), methane (CH_4), particulate matter (PM), and carbon dioxide (CO_2). Studies aimed at identifying the benefits of a flower roundabout in respect of standard intersections have been carried out in terms of road pollution emission, together with specific traffic analyses. These studies relate to six different geometric layouts and many traffic flow conditions [9].

As stated by Corriere et al. [8], when the traffic intensity is high (up to 450,000 veh/year), road emissions are a function of the roundabout's geometry; pollutant emissions have been calculated by considering the annual cumulative pollutants correlated only to the daily hourly peak flow, in the working days (5 days/week). In this case, two-lane roundabouts provide better performances; meanwhile flower roundabouts provide intermediate performances between conventional roundabouts with (1 + 1) or (1 + 2) geometry and two-lane roundabouts. Only when the right-hand turning percentage is higher than 70 % of the total, flower roundabouts can cause delays and pollutant emissions inferior to those observed within the other configurations examined.

5.3 Dual One-Lane Roundabouts on Two Levels with Right-Hand Turning Bypasses—the "Target Roundabout"

The "target roundabout" is presently also at the development phase. A target roundabout [10] is designed as a two one-lane roundabout with different outer diameters, located on dual levels (see Fig. 5.6), and all right-hand turners on both roundabouts have their own, separate right-hand turn bypass lanes. The dual one lane roundabout on two levels (see Fig. 5.7) allows driving from all directions to all directions, and it also "forgives errors"; if a driver mistakenly stays on the left-hand lane at the entrance it is still possible to turn right at the next exit (different to the turbo-roundabout).

Driving at a target roundabout is the same as on the turbo-roundabout (the same philosophy of signposting and lane-marking).

One of the basic characteristics of the target roundabout is the same as at the turbo and flower roundabouts—physically separated traffic lanes within a circulatory carriageway; bypasses and one-lane circulatory roadway sections. All right-hand turners have their own separated traffic lanes; consequently the inner circulatory roadway is used only by vehicles that drive through a roundabout (180°), turn for three quarters of a circle (270°), or turn semicircle (360°).

By physically separating the right-hand turning traffic flow, two one-lane roundabouts are obtained, with no crossing conflict spots (unlike in the case of the standard two-lane or turbo-roundabout), and also no weaving conflict spots (unlike in the case of the standard two-lane roundabout). Any possible weaving conflict spots when

Fig. 5.6 Typical layout of a target roundabout; sketch

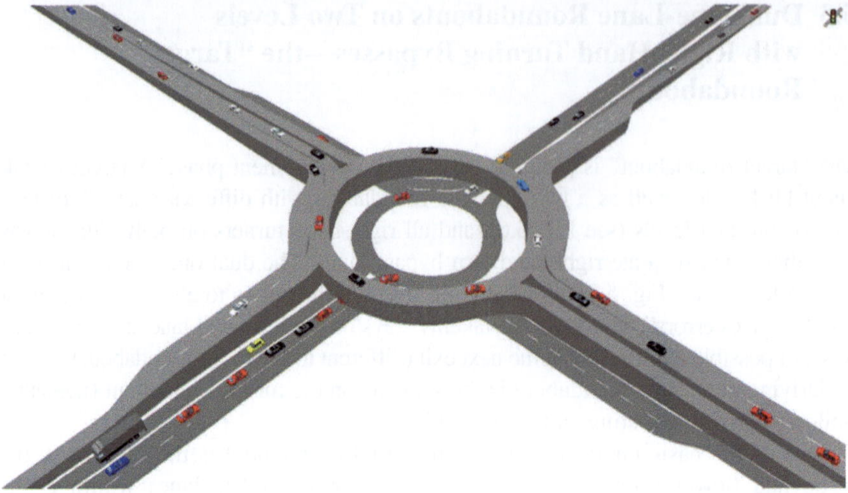

Fig. 5.7 3D rendering of a target roundabout

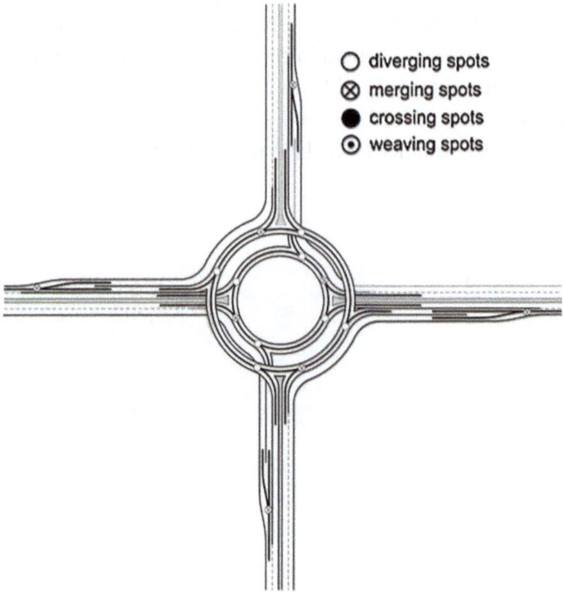

Fig. 5.8 Conflict spots at a target roundabout

transferring from the circulatory carriageway onto the road section are in front of a roundabout (as in the case of the turbo and flower roundabouts), being a safer solution from the traffic-safety point of view. At the target roundabout there are just 8 merging and 8 diverging conflict spots (as at the two one-lane roundabouts) (Fig. 5.8).

A target roundabout is especially useful within suburban areas, with plenty of space, where two-level interchanges (standard diamond, diverging diamond, cloverleaf interchange…) are all possible solutions. However, this solution is acceptable also in urban areas due to small size.

In accordance with the results [10] of the micro simulation, carried out using VISSIM, we can summarize that the target roundabout (with diameter of the larger roundabout D = 75 m) could serve the interchange with 50,000 AADT with a good level of service and up to 60,000 AADT with a level of service F, in accordance with HCM 2010 criteria. Compared with e.g. the cloverleaf interchange, this would be a big disadvantage due to capacity criteria, but in the case of urban space limitation, the possible target roundabout would need to be taken into consideration and analyzed using forecasted traffic.

5.4 Roundabout with Segregated Left-Hand Turning Slip-Lanes on Major Roads—the "Four Flyover Roundabout"

The roundabout with segregated left-hand turning bypasses (slip-lanes) on major roads—in short the "four flyover roundabout" (Fig. 5.9) is designed as a one large one-lane roundabout at upper, and both left-hand turners on the major roads have their own separate left-hand turn bypass lanes, located at another, lower level. Left-hand turners are located as on standard intersections—at the left lane on the approach (see Fig. 5.10).

By physically separating left-hand turning traffic flow on major roads, we obtain a one-lane roundabout by physically separating the left-hand turning traffic flow on major roads, with no crossing and also no weaving conflict spots. Any possible

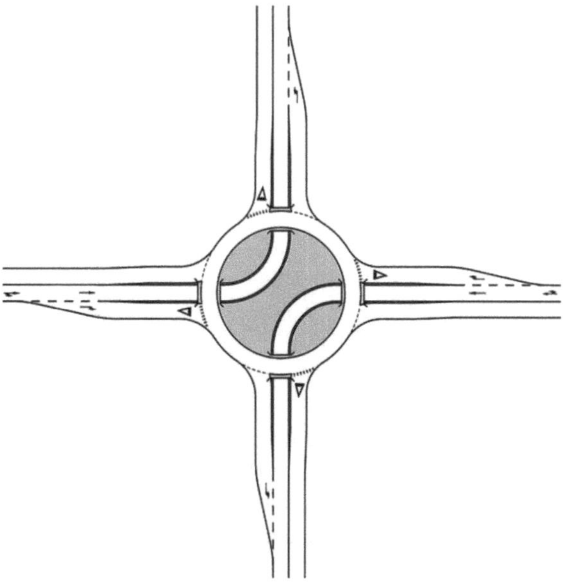

Fig. 5.9 A roundabout with segregated left-hand turning slip-lanes on major roads—the "four flyover roundabout"; sketch

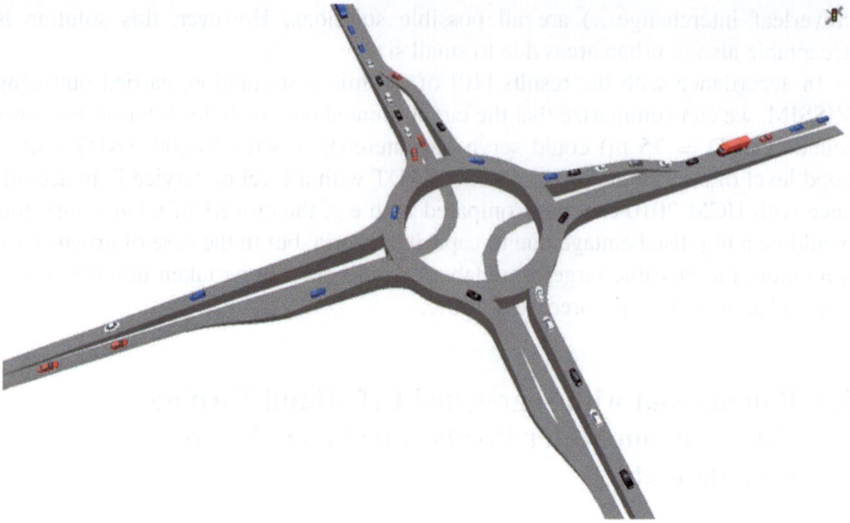

Fig. 5.10 3D rendering of a four flyover roundabout

Fig. 5.11 Conflict spots at a four flyover roundabout

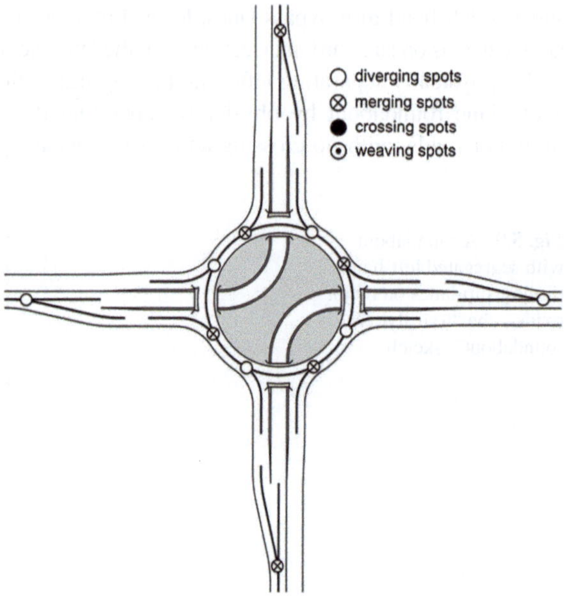

weaving conflict spots when transferring from the circulatory carriageway onto the road section are in front of a roundabout (as in the case of the turbo, flower, and target roundabouts), being a safer solution from the traffic-safety point of view.

At a four flyover roundabout there are just 6 merging and 6 diverging conflict spots (see Fig. 5.11).

A four flyover roundabout is especially useful in urban areas, where we do not usually have plenty of space, and standard two-level interchanges (standard diamond, diverging diamond, cloverleaf interchange…) are usually not feasible solutions.

Following the results of micro simulation, it could be summarized that four flyover roundabout (with diameter D = 90 m) could serve an interchange with about 45,000 AADT.

There are few variations of this type of solution. It could be constructed as a standard one-lane roundabout or with segregated right-hand turning slip-lanes. One of variations is presented in the continuation.

5.5 Roundabout with Segregated Left-Hand Turning Slip-Lanes on Major Roads and Right-Hand Turning Slip-Lanes on Minor Roads—the "Roundabout with Left and Right Slip-Lanes"

A roundabout with segregated left-hand turning slip-lanes on major roads and right-hand turning slip-lanes on minor roads, in short the "roundabout with left and right slip-lanes" (Fig. 5.12), is a variation of the four flyover roundabout, and is also at this moment at the development phase. A roundabout with left and right slip-lanes is also a hybrid like a four flyover roundabout, combining a one-lane roundabout, right-hand turning slip-lanes, and left-hand turning slip-lanes located on another, lower level.

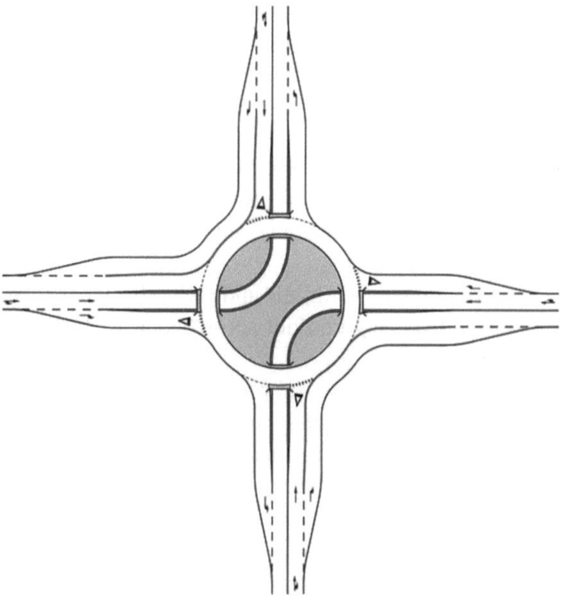

Fig. 5.12 Typical layout of a roundabout with segregated left-hand turning slip-lanes on major roads and right-hand turning slip-lanes on minor roads—the "roundabout with left and right slip-lanes"; sketch

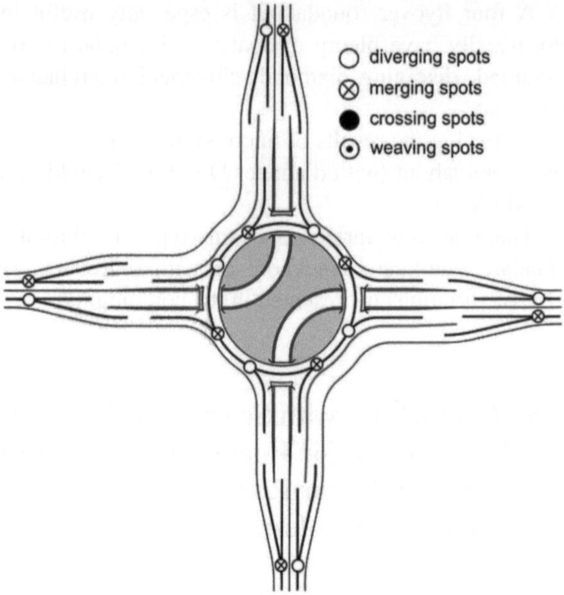

Fig. 5.13 Conflict spots at a roundabout with left and right slip-lanes

This type of roundabout is designed as one large one-lane roundabout at one level, and both left-hand turners on major roads have their own separate left-hand turn bypass lanes, located on another, lower level. On minor roads, bypasses for a right-hand turning are added. Left-hand turning lanes are located as on a standard intersection—at the left lane on the approach, and the right-hand turning slip-lanes are constructed as at a turbo or flower roundabout (divided curbs).

This type of roundabout also allows driving from all directions to all directions, and it also "forgives errors". At this type of roundabout there are no crossing conflict spots (unlike in the case of the turbo-roundabout), and also no weaving conflict spots (unlike in the case of the standard two-lane roundabout). Any possible weaving conflict spots when transferring from the circulatory carriageway onto the road section are in front of a roundabout (as in the case of the turbo, flower, target and four flyover roundabout), surely being a safer solution from the traffic-safety point of view.

At a roundabout with left and right slip-lanes (Fig. 5.13) there are just 8 merging, and 8 diverging conflict spots (as at the two one-lane roundabouts and target roundabout).

All alternative types of roundabouts have their advantages and deficiencies, which makes sense, as they are intended to solve particular problems. In the near future, we can expect further developments of alternative types of roundabouts, intended for solving specific problems, which will certainly represent a challenge to our branch of science.

References

1. Tollazzi, T., Renčelj, M., & Turnšek, S. (2011). New type of roundabout: Roundabout with "depressed" lanes for right turning—"flower roundabout", *Promet—Traffic & Transportation, Scientific Journal on Traffic and Transportation Research, 23*(5). doi: 10.7307/ptt.v23i5.153.
2. Silva, A. B., Santos, S., & Gaspar, M. (2013). Turbo-roundabout use and design. In *CITTA 6th annual conference on planning research, responsive transports for smart mobility*, Coimbra, Portugal.
3. Yap, Y. H., Gibson, H. M., & Waterson, B. J. (2013). An international review of roundabout capacity modelling. *Transport Reviews: A Transnational Transdisciplinary Journal, 33*(5). doi:10.1080/01441647.2013.830160.
4. Mauro, R., & Guerrieri, M. (2013). Flower roundabouts: performances analysis and comparison with conventional layouts. *European Journal of Scientific Research, 94*(2), 242–252.
5. Mauro, R., & Guerrieri, M. (2012). Right-turn bypass lanes at roundabouts: Geometric schemes and functional analysis. *Modern Applied Science, Canadian Center of Science and Education, 7*(1), 1–12. doi:10.5539/mas.v7n1p1.
6. Guerrieri, M., Ticali, D., Corriere, F., & Galatioto, F. (2012). Flower roundabouts: Estimation of capacity and level of service. *GSTF Journal On Computing (Joc), 2*(2), 101–106. doi:10.5176/2010-283_2.2.175.
7. Mauro, R., & Cattani, M. (2004). Model to evaluate potential accident rate at roundabouts. *Journal of Transportation Engineering, 130*(5), 602–609. doi:10.1061/(ASCE)0733-947X(2004)130:5(602).
8. Corriere, F., Guerrieri, M., Ticali, D., & Messineo, A. (2013). Estimation of air pollutant emissions in flower roundabouts and in conventional roundabouts. *Archives of Civil Engineering, 59*(2), 229–246. doi:10.2478/ace-2013-0012.
9. Mauro, R., & Cattani, M. (2012). Functional and economic evaluations for choosing road intersection layout. *Promet—Traffic & Transportation, Scientific Journal on Traffic and Transportation Research, 24*(5). doi:http://dx.doi.org/10.7307/ptt.v24i5.1180.
10. Tollazzi, T., Jovanović, G., & Renčelj, M. (2013). New type of roundabout: Dual one-lane roundabouts on two levels with right-hand turning bypasses—"Target roundabout", *Promet—Traffic & Transportation, Scientific Journal on Traffic and Transportation Research, 25*(5). doi:10.7307/ptt.v25i5.1230.

Chapter 6
General Criteria for Calculating the Capacities of Alternative Types of Roundabouts

6.1 Introduction

Roundabouts are an increasingly common form of junction worldwide and their effective design requires detailed analysis of maximum vehicle throughput capacities. Since the 1970s, a series of models have been developed worldwide for an estimation of the capacity of roundabouts, almost all of which have relied upon extensive empirical data due to the complexity of the physical and behavioral processes affecting roundabout entry capacities. However, given the different fundamental principles (and particularly the geographical origins) of models, it is important to have a clear understanding of their limitations and their applicability within new contexts. These models are based on three main methodologies: empirical, gap acceptance and simulation. Due to their limitations, each of these methodologies on their own cannot completely explain the complex behavioral and physical processes involved at roundabout entries, hence all models require strong semi-empirical or fully-empirical bases using data obtained from their countries of origin.

The point of this book is not capacity calculation of roundabouts, and the point of this chapter is not describe the worldwide state-of-the-art on this field, but just to present a short overview of major capacity models, their limitations, and the capacity calculation of the alternative types of roundabouts, the recent and those at the development phases.

6.2 Roundabout Capacity

Delays on roundabouts consist of geometric and queuing delays. The latter arise from queues resulting from a combination of random arrivals and oversaturated conditions, and are typically estimated using time-dependent queuing models [1] such as those of Akçelik et al. [2]. Other methods to determine queues and delays

T. Tollazzi, *Alternative Types of Roundabouts*, Springer Tracts on Transportation and Traffic 6, DOI 10.1007/978-3-319-09084-9_6

include those based on equivalent blocked/unblocked periods in gap acceptance for back-of-queue estimation in SIDRA [3] or those in microscopic simulation models [4]; however, as a rule, greater capacity leads to smaller queues and delays.

The entry capacity can be defined as the maximum inflow from a roundabout entry with saturated demand, where at least one vehicle is always queued at the give-way line of the entry lane ready to enter any available acceptable gap in a circulating stream. This flow is averaged over the applicable analysis time interval to account for inherent short-term (i.e. minute-by-minute or vehicle-by-vehicle) variability resulting from the gap acceptance process [1]. With the offside-priority rule, the entry capacity varies according the prevailing circulating flow across the entry, and also depends on the geometry. For example, a wider multi-lane entry enables more than one vehicle to enter the same available gap, whilst for example slip-lanes increase capacity for traffic turning towards the first arm downstream [5, 6].

The slip-lanes can be distinguished according to the layout and the entry control type (stop, yield or free-flow acceleration lane). Theoretical and empirical models for calculation of slip-lanes capacity were carried out by Tracz [7] and Tracz et al. [8], Al-Gandur et al. [9] and Mauro and Guerrieri [5].

Capacity has also been found to be affected by environmental factors including snow, ice, and rain, as well as other traffic factors aside from circulating flow. Pedestrian crossings (Fig. 6.1) also reduce entry capacity, either by interrupting demand flows at the entry or causing queues within the circulatory carriageway. The effect of pedestrian flow on entry capacity can be evaluated by means of the German method [10], the Marlow and Maycock formula [11] or by means of the CETE de l'Ouest Formula [12]. Roundabout capacity can be considered as a function of the geometry and demand flows, as well as driver and vehicle characteristics. For example, heavy vehicles reduce the entry and ring lanes capacity; for this reason, the flow rate for each movement may be adjusted to account for vehicle stream characteristics using specific factors (Passenger Car Equivalent factor) [13]. Consequently, a large number of factors and variables need to be assumed in order influence the gap acceptance process and capacity.

Entry capacity can be modelled by lane or by approach, where for a multilane entry, an approach or an arm capacity is usually not a sum of the individual lane capacities. The entry capacity is a sum of the individual lane capacity only in the case of an equal degree of saturation. For example if C_1 and C_2 are the capacities of lanes 1 and 2 at the entry:

if $x_1 = x_2$: $C_E = C_1 + C_2$
where:
$x_1 = Q_1/C_1$
$x_2 = Q_2/C_2$
Q_1 and Q_2 are the entry flows at lane 1 and 2 (see [14] and [15]);
if $x_1 \neq x_2$: $C_E = (Q_1 + Q_2)/\max(x_1; x_2)$
(see [15, 16]).

The circulating flow across an entry (and thus the entry's capacity) depends on the junction turning movements and the demand flows and capacities of the other arms at the roundabout. Hence, iterative methods are used to determine the final

Fig. 6.1 Pedestrian crossing reduces entry capacity

entry capacities for all arms for a given set of design flows [1]. Consequently, several viable capacity models worldwide have thus been developed, which can be classified by their primary methodologies into the following categories:

- empirical models: based on relationships between the geometry and the actual measured capacity;
- gap-acceptance models: based on understanding driver's behavior;
- micro-simulation models: based on the modelling of vehicle kinematics and interactions.

In the continuation, just a brief and concentrated explanation is presented, based on analyses of the multi-lane roundabouts.

For the exits and ring capacity, filed observations show that the exit capacity limit for each lane is in the range of 1,200–1,400 veh/h, instead the ring capacity is correlated to the number of the entry and circulating lanes and to roundabouts typology (mini, medium and large). Therefore, the ring capacity range is 1,600–2,500 veh/h [17].

6.2.1 Empirical Models

Empirical capacity models, based on the calibration of relationships between geometry and actual measured capacity, are the longest established forms. Empirical regression models are created through statistical multivariate regression analyses to fit mathematical relationships between measured entry capacities (Q_e), circulating flows (Q_c) and other independent variables which significantly affect entry capacity, such as geometric elements of a roundabout [18].

The relationship between Q_e and Q_c is usually assumed to be linear ($Q_e = \alpha - \beta Q_c$) or exponential ($Q_e = \alpha e^{-\beta Q_c}$). Entry capacity can be directly measured from observed entry flows during continuous queuing at the entry, which are typically recorded with the corresponding circulating flows over time intervals of 0.5, 1 or more minutes [1].

Several linear regression models are widely-acknowledged to be the best examples of fully-empirical roundabout capacity models.

Empirical capacity models have their limitations as they map the relationship between input parameters and capacity, but do not necessarily prove causality nor provide a complete theoretical understanding of those relationships. Although this does not obviate their uses as predictive tools, it is important to understand the underlying principles as there may be atypical scenarios where engineering judgment is needed to assess the validities of the predicted capacities. This is a particular issue with roundabout design, which may need to conform to unusual site constraints with different arm sizes or orientations. Many empirical models are likely to have been constrained by the sample sizes used for model development, which would have been limited by the number of congested roundabout entries available. Statistically-significant relationships between capacity and geometric parameters could also have been difficult to identify due to the limited range of observable parameter values. The results of any empirical model are also likely to be reliable only within the range of parameters within the original database used to develop it [1].

6.2.2 Gap-Acceptance Models

Gap-acceptance is an alternative approach to modelling capacity, based on theoretical models developed around parameters obtained from the measurements of individual headways between circulating and entering vehicles. The data collection for this method is thus less contingent on heavily-congested entries with continuous queuing compared to those of empirical models [1]. Gap-acceptance models rely on three variables for determining entry capacity:

- critical gap (t_c) is the minimum time headway in the circulating stream that an entering driver will accept. As it cannot be observed directly, many methods have been developed for its estimation from observed rejected and accepted gaps;
- follow-on headway (t_f) is the time headway between two consecutive queued vehicles entering the same gap within the circulating stream;
- distribution of gaps within the circulating flow is based on Poisson random arrivals or bunched flows.

Several gap-acceptance models are widely-known nowadays [19–22] but it seems that one of the best-known gap-acceptance model for roundabouts was developed in Australia, the SIDRA model [1]. The latest SIDRA model [23] was a further development of the Troutbeck's SR45 model using a traffic signal analogy and revised versions of the empirical follow-on headway and critical gap equations from SR45. Other revisions to the circulating headway and capacity models have included additional factors for priority-sharing, origin-destination patterns, and queuing on upstream approaches [1].

Another computer program for the capacity calculation of conventional and also alternative types of roundabouts is the KREISEL software [24, 25]. It seems that at the moment, KREISEL (based on closed-form equations) is, with the exception of VISSIM, the only software that allows the evaluating of capacity, delays, and queues at turbo-roundabouts.

Gap-acceptance models also have their limitations. One criticism of gap-acceptance based models is that they do not directly quantify the relationship between geometry (the only factor which can be controlled by a roundabout designer) and capacity. Instead, they require the formulation and calibration of an intermediary vehicle-vehicle interaction model, which then has to be related separately to geometry and entry capacity. This is an issue as capacity models are sensitive to values of critical gap and follow-on headway, as well as differences in headway distributions at higher circulating flows; there are also difficulties when defining the parameters from field-measurements [1].

6.2.3 Micro Simulation Models

Traffic simulation is increasingly being used to assess traffic operations along many different types of roadway networks. From highways to arterial streets, traffic simulation enables the engineer as well as the public to visualize traffic operations. Simulation methods can be generally divided into two main groups; macroscopic, and microscopic models. Macroscopic models combine vehicles and travelling amongst groups, the traffic flow is presented as a statistical model; results are presented as an average value after a certain time. With macroscopic models the emphasis is placed on the links and the intersections are simplified in the model. Unlike microscopic models, macroscopic models focus on a long-term planning period. With microscopic models every vehicle, pedestrian, or cyclist can be described by their real characteristics (dimensions, speed, accelerations, decelerations, etc.). Microscopic models are usually used for traffic flow analyses within a short-term planning period.

As pointed out above, microscopic simulation models are based on modeling movements and interactions of individual vehicles with their real characteristics, on a network consisting of links and nodes or connectors. Vehicle movements are governed by the gap acceptance, car-following, lane-changing and other models, and are typically calculated for each vehicle at every specified time-step. Driver behavior parameters such as critical gaps, and processes such as vehicle generation are stochastically assigned through Monte Carlo methods using specified

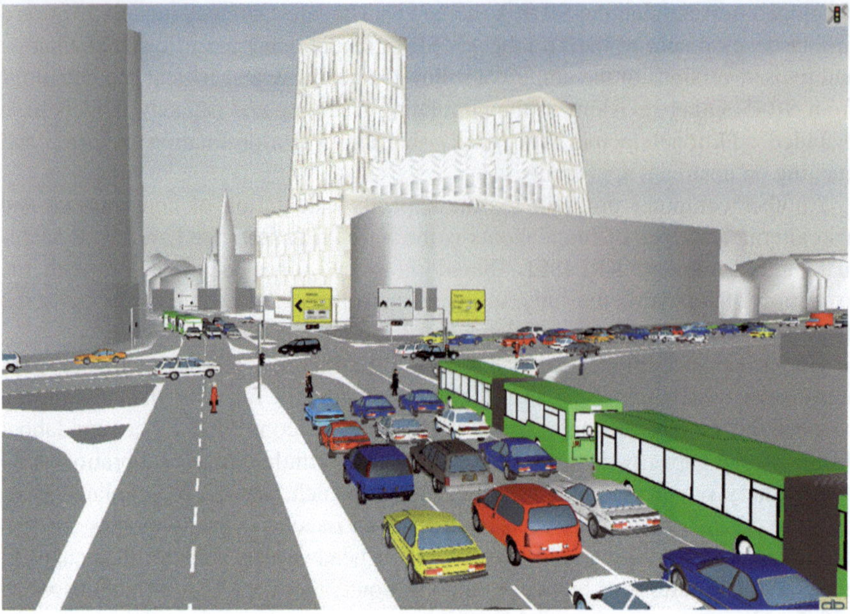

Fig. 6.2 3D presentation of a simulated urban intersection including pedestrians and cyclists

probability distributions; the resulting variability of outputs attempt to reflect the characteristics of real-world traffic.

Several proprietary microscopic simulation programs are available for the modelling of general traffic networks. However, the microscopic traffic flow simulator VISSIM has proved to be an extremely appropriate tool for capacity calculations during the process of creating pilot projects [26]. VISSIM is a stochastic, discrete, time-oriented microscopic simulation model, offering a wide variety of urban and highway applications for integrating public and private transportation. Complex traffic conditions are visualized at a high level of detail supported by realistic traffic models.

Vehicle movements may be animated in 2D or in 3D (Fig. 6.2). This feature allows users to create realistic video clips in AVI format, which can be used for communicating a project's vision. Background mapping capabilities with aerial photographs and CAD drawings should be applied for better representation.

It uses psycho-physical characteristics of the so-called "car following" model for the longitudinal motions of vehicles and algorithms, based on the rules of driving for vehicles, coming from side-on directions. VISSIM is a commercial software tool with about 7,000 licenses distributed worldwide in the last 15 years. About one-third of users are within consultancies and industry, one-third within public agencies, and the remaining third are applied at academic institutions for teaching and research, but the software is primarily suitable for traffic engineers. However, as transport planning is looking toward a greater level of detail, an increasing number of transport planners use micro simulation as well. Traffic engineers and transport planners assemble applications by selecting appropriate objects from a variety of primary building blocks. In order to simulate multi-modal traffic flows, the technical features of pedestrians, cyclists, motorcycles, cars, trucks, buses, trams, and railways, are provided with the options of customization [26].

The mathematical model was conceived on an idea resulting from the Wiedermann theory. The basic idea of the model is presented through the presumption that any driver can experience one of the following situations:

- driving in the free traffic flow (without the influence of other vehicles);
- approach-driving (the process of adjusting the speed to the speed of the vehicle ahead);
- follow-driving (the driver maintains a constant distance between the vehicle ahead, without accelerating or braking);
- braking (applied, when the safety distance drops below the lower level).

Typical for the VISSIM is that it does not use the conventional link/node mode modelling system but rather the link/connector system that enables a designer to model extremely complex intersection geometries. Based on the digital aerial-photo shot (DOF) and the segment from the reconstruction or construction project, the mathematical model of the normal and/or turbo roundabout is created in the first step, by series of links (links and connectors).

The calibration of the microscopic model follows in the next step. The VISSIM includes a series of simulation parameters that can influence the simulation results (characteristics of network, vehicles, drivers), whereas in the calibration process we must focus on parameters, that are defined within the so-called Priority Rules within VISSIM. In regard to parameters, the rules of driving are determined as well as the minimal critical time (drivers' reaction time) and minimal distance (headway). VISSIM determines which traffic participants have priority by setting up Priority Rules. Depending on the conditions at the conflict area, an individual decides whether to continue the path or wait for appropriate traffic conditions. On the marked spot, driver must always check both of the predetermined conditions [minimal distance (minimal headway) and minimal critical time (gap time)], before continuing the path. More marked spots (conflict marker-green) can belong to one stop line (stop line-red).

Different critical times (gap times) are used for different categories of vehicles. In the model, we must consider the fact that the inner roundabout lane becomes much more attractive for users. For the correct analysis and evaluation of capacity parameters, we must collect traffic characteristics and results of the real-time simulation. It is sensible to evaluate the success of the analyzed geometry based on the following criteria: average delay per vehicle(s) (considering all types of vehicles), average catchment length (m) and maximum catchment length (m). Thereby it is very important that during the micro simulation, we can visually observe the running of traffic flows.

Another traffic micro simulation software widely used around the world is AIMSUN, that allows for estimating the capacity, delay and other Measure Of Effectiveness (MOE) at roundabouts. AIMSUN is a widely used commercial transport modeling software, developed and marketed by TSS—Transport Simulation Systems based in Barcelona, Spain. Microscopic and mesoscopic simulators are the components of AIMSUN that allow for dynamic simulations. They can deal with different traffic networks: urban networks, freeways, highways, ring roads, arterials, and any combination thereof. This software is used to improve road infrastructure, reduce emissions, cut congestion and design urban environments for vehicles and pedestrians. AIMSUN stands out for the exceptionally high speed of its simulations and for fusing travel demand modelling, static and dynamic traffic assignment with mesoscopic, microscopic and hybrid simulations—all within a single software application. The input data required by AIMSUN Dynamic simulators is a simulation scenario, and a set of simulation parameters that define the experiment [27].

However, one of the main advantages of microscopic simulation models is that demand flows and turning movements can be controlled for parametric studies. They are thus used in roundabout research, which requires such effects to be modelled. The most widely-acknowledged limitation regarding the microscopic simulation modelling of roundabouts is the priority-reversal and priority-sharing phenomena. Whilst the former may arise due to the capacity restrictions of other junctions downstream and is thus beyond the scope of this book, the more subtle issue of priority-sharing, which occurs especially at high circulating flows, does need to be considered. The outputs of microscopic simulation models also depend on a large number of different parameters that govern a vehicle's movements. Many of these parameters can be difficult to calibrate from available field data, and so may be left as default values recommended by the software developers. Calibration and validation of the models are thus crucial to ensure the suitability of these parameters [1].

The limitations of each of the major capacity modelling approaches, as briefly presented above, mean that the development of the major capacity prediction models has usually involved a combination of two or more approaches. As pointed-out, the major component of capacity model errors arises from the variability of the driver and traffic behavior in the gap acceptance process [1], as reflected by the poorer capacity predictions for junctions with gap-acceptance relative to those for traffic signal controlled junctions, and differences in capacities between different countries have also been attributed to different periods of roundabout experience [17]. However, the differences between the models' outputs and observed conditions can be reduced through calibration, and several studies have applied recommended calibration methods to compare the models against hypothetical or actual data.

6.3 Capacity Calculation of the Alternative Types of Roundabouts

As pointed-out previously, today after many years of experience, there are different ideas about the "ideal roundabout", with little consensus on the crucial effect of rules regarding how to negotiate an intersection. There are several different types of roundabouts worldwide today, called the alternative types of roundabouts. Some of them are already in frequent use all over the world (hamburger, dumb-bell…), some of them are recent and have only been implemented within certain countries (turbo, dog-bone, compact semi-two-lane circle…), and some of them are still at the development phase (turbo-square, flower, target, with segregated left-turn slip-lanes…). Alternative types of roundabouts typically differ from "standard" one- or two-lane roundabouts in one or more design elements, as their purposes for implementation are also specific. The main reasons for their implementation are the particular disadvantages of "standard" one- or two-lane roundabouts regarding actual specific circumstances. Usually, these disadvantages are highlighted by low-levels of traffic safety or capacities.

Two types of analytical models are used for the capacity calculation of multilane roundabouts (as alternative types of roundabouts usually are). The first is based on the traffic flow theory, where it is presumed that the capacity of the entry depends on the intensity of the circulating traffic flow and potential conflict traffic flow, immediately before the exit from the roundabout. Depending on the used model/equation, the ratio between the capacity of the entry and the intensity of the circulating traffic flow can be linear or exponent, whilst the capacity depends on the geometrical characteristics of the roundabout.

The second type of analytical models for the capacity calculation of multilane roundabouts is based on the theory of time gaps in the traffic flow, where the interaction of two traffic flows is monitored. The capacity of the entry is determined based on the available time gaps in the circulatory traffic flow and utilization of these time gaps by the entry traffic flow. Parameters, which determine this interaction, are: minimum time gap [minimum distance (headway)] in the circulating traffic flow, average waiting time at the entry and critical time gap [minimum critical time (gap time)].

As recent alternative types of roundabouts are relatively new, practical evaluation data are presently unavailable for them. A slightly worse situation occurs in the cases of alternative types of roundabouts presently at the development phase, because they do not as yet exist. Therefore, different possibilities remain open for determining the capacities of these types of roundabouts.

6.3.1 Capacity Calculation of the Recent Alternative Types of Roundabouts

As pointed above, practical evaluation data are presently unavailable for the recent alternative types of roundabouts. For example, only in The Netherlands have a number of turbo-roundabouts been constructed and very few of those are operating at/or near capacity. Therefore, different possibilities remain open for determining the capacities of turbo-roundabouts and even other recent alternative types of roundabouts.

In The Netherlands, the modified Bovy equation for the capacity calculation of the turbo-roundabout was chosen because of two main reasons, it:

- includes the influence of the "left-turning traffic flow";
- enables simple modelling of the traffic distribution on traffic lanes.

It is because change of traffic lanes is impossible in turbo-roundabouts, that the traffic distribution is clearly defined. Correction factors regarding the influence of the exiting traffic flow "a" (α in the original equation) and the influence of the circulating traffic flow are different for two circulatory traffic lanes in the circulatory carriageway. Fortuijn [28, 29] modified the equation so that he divided the β parameter to b_1 (inner circulatory traffic lane) and b_2 (outer circulatory traffic lane). In this way, each traffic lane can be entered into the calculation separately. Depending on the distribution of intensity by lanes, b_1 and b_2 may have minimum or maximum values.

Despite its simplicity, the Bovy equation produces correct results. The equation is considered to be simple because it is based on the roundabout geometry and linear distribution of results.

The calculation of capacity regarding turbo-roundabouts according to the latest Dutch guidelines, is based on the modified Bovy equation, which was also the basis for creating the computer program "Explorer of multilane roundabouts" which is an integral part of the Dutch guidelines [30]. Some other capacity models have already been created for turbo- and other recent alternative types of roundabouts [9, 14, 15], VISSIM plays a significant role as well because of "lane selection" and "lane changing" [26].

The most recent models show that entry capacity at turbo-roundabouts is conditioned by the individual lane capacity, by conflicting vehicles and pedestrian flow, by the combination of circulating flows along lanes at the ring, by user's behavior (minimum gap time, minimum headway), as well as by the balance of traffic flows at the entry [15, 31]. Therefore, contrary to models for conventional roundabouts, at entries of turbo-roundabouts there is no biunique relation between circulating flow and entry capacity but a continuous set of capacity values are related to lanes degrees of utilization.

6.3.2 Capacity Calculation of the Alternative Types of Roundabouts at Development Phases

A slightly different situation applies in the case of alternative types of roundabouts at the development phases because they do not as yet exist, and it is impossible to presume and calibrate any of the influenced factors. As alternative types of roundabouts at the development phases are practically unique, there are not specific program tools for them. Accordingly, software tools that allow the designer to prepare unique solutions are advantages.

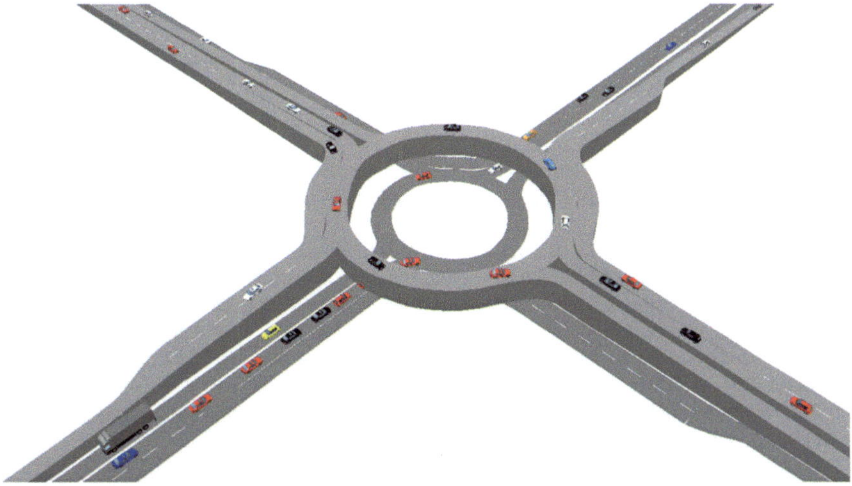

Fig. 6.3 3D presentation of a simulated target roundabout

It seems that at the moment the best approach to the capacity calculations of alternative types of roundabouts at the development phases is a comparative analysis of the different types of multilane roundabouts using a PTV VISSIM micro-simulation program (Fig. 6.3). As we cannot trust the absolute values of results, at least a comparison between different types of roundabouts can be carried out (e.g. flower or target roundabouts compared with a standard two-lane roundabout) [32, 33].

Instead of testing and calibrating the positions and parameters (minimum gap time, minimum headway) for roundabouts at the development phases, it is appropriate to use the standard parameters that have been tried and tested under local (country) conditions.

References

1. Yap, Y. H., Gibson, H. M., & Waterson, B. J. (2013). An international review of roundabout capacity modelling. *Transport Reviews: A Transnational Transdisciplinary Journal, 33*(5). doi:10.1080/01441647.2013.830160.
2. Akçelik, R., Chung, E., & Besley, M. (1998). *Roundabouts: Capacity and performance analysis*. Research Report ARR 321, ARRB Transport Research, Vermont South, Australia.
3. Akçelik, R. (1994). Gap acceptance modelling by traffic signal analogy. *Traffic Engineering and Control, 35*(9), 498–506.
4. Lanović, Z. (2005). Selection of traffic flow optimization model. In *ICTS 2005 transportation logistics in science and practice, 9th international conference on traffic science*. Portorož, Slovenija, November 14–15, 2005.
5. Mauro, R., & Guerrieri, M. (2013). Right-turn bypass lanes at roundabouts: Geometric schemes and functional analysis. *Modern Applied Science, Canadian Center of Science and Education, 7*(1), 1–12. doi:10.5539/mas.v7n1p1.

6. Tollazzi, T., Renčelj, M., & Turnšek, S. (2011). New type of roundabout: roundabout with "depressed" lanes for right turning—"flower roundabout". *Promet—Traffic and Transportation, Scientific Journal on Traffic and Transportation Research, 23*(5), 353–358. doi:10.7307/ptt.v23i5.153.
7. Tracz, M. (2008). Analysis of small roundabouts' capacity. In *National roundabout conference, TRB.* Kansas City, Missouri, USA.
8. Tracz, M., Chodur, J., & Ostrowski, K. (2011, June) Roundabouts country report—Poland. In *6th International symposium on highway capacity and quality of service.* Stockholm, Sweden.
9. Al-Ghandour, M., Schroeder, B., Rasdorf, W., & Williams, B. (2012). Delay analysis of single-lane roundabout with a slip lane under varying exit types, experimental balanced traffic volumes, and pedestrians: Using microsimulation. *TRB*, 91st Annual Meeting, January 22–26, 2012.
10. Brilon, W., Stuwe, B., & Drews, O. (1993). *Sicherheit und Leistungsfähigkeit von Kreisverkehrsplätzen.* Bochum Deutschland: Institute for Traffic Engineering, Ruhr Universität.
11. Marlow, M., & Maycock, G. (1982). *The effect of zebra crossing on junction entry capacities*, Report SR 742, Crowthorne, Berkshire, England: Transport and road research laboratory (TRRL).
12. Louah, G. (1992). Panorama critique des modeles Français de capacité des carrefours giratories, Actes du séminaire international "Giratoris 92", Nantes, France, October 14–16, 1992.
13. Rodegerdts, L., Bansen, J., Tiesler, C., Knudsen, J., Myers, E., Johnsonm, M., et al. (2010). NCHRP Report 672, *Roundabouts: An Informational Guide* (2nd ed.). Washington, DC: Transportation Research Board.
14. Mauro, R., & Guerrieri, M. (2013). Right-turn bypass lanes at roundabouts: Geometric schemes and functional analysis. *Modern Applied Science*, *Canadian Center of Science and Education, 7*(1). doi:10.5539/mas.v7n1p1.
15. Mauro, R., & Branco, F. (2010, April). Comparative analysis of compact multilane roundabouts and turbo-roundabouts. *Journal of Transportation Engineering*, *136*(4). doi:10.1061/(ASCE)TE.1943-5436.0000106.
16. Corriere, F., Guerrieri, M., Ticali, D., & Messineo, A. (2013). Estimation of air pollutant emissions in flower roundabouts and in conventional roundabouts. *Archives of Civil Engineering, 59*(2), 229–246. doi:10.2478/ace-2013-0012.
17. Mauro, R. (2010). *Calculation of roundabouts: Capacity, waiting phenomena and reliability.* Berlin: Springer.
18. Guichet, B. (2005). Roundabouts in France and new use. In *National roundabout conference, TRB*, Vail, Colorado, USA, May 22–25, 2005.
19. Transportation Research Board. (2010). *Highway capacity manual 2010,* Washington, DC.
20. Brilon, W., & Stuwe, B. (1993). *Capacity and design of roundabouts in Germany*, Transportation Research Record.
21. Brilon, W., Wu, N., & Bondzio, L. (1997). Unsignalized intersections in Germany—A state of the art 1997. In *Proceedings of the third international symposium on intersections without traffic signals* (pp. 61–70). Portland, Oregon, USA, July 21–23, 1997.
22. Wu, N. (2001). A universal procedure for capacity determination at unsignalized (priority-controlled) intersections. *Transportation Research Part B: Methodological*, *35*(6), 593–623. doi:10.1016/S0191-2615(00)00012-6.
23. *SIDRA.* (2013). *Intersection 6.0.* Greythorn, Victoria: Akcelik & Associates Pty Ltd.
24. Brilon, W. (2005). Roundabouts: A state of the art in Germany. In *National roundabout conference, TRB.* Vail, Colorado, USA, May 22–25, 2005.
25. Trentino Mobilità S. p. A. (2014). *KREISEL: Software per la valutazione della capacità e della qualità del flusso nelle rotatorie.* Accessed April 7, 2014, from http://www.trentinomobilita.it/rotatorie.html.
26. Fellendorf, M. & Vortisch, P. (2010). Chapter 2: Microscopic traffic flow simulator vissim. In J. Barceló (Ed.), *Fundamentals of traffic simulation, international series in operations research and management science 145.* Springer Science+Business Media, LLC. doi: 10.1007/978-1-4419-6142-6_2.

27. Barceló, J. (2014). AIMSUN microscopic traffic simulator: A tool for the analysis and assesment of ITS systems. *TSS-Transport Simulation Systems, Technical Notes on GETRAM/AIMSUN*. Accessed April 7, 2014, from http://www.polloco.pl/events/pdf/j_barcelo_3_1.pdf.
28. Fortuijn, L. G. H. (2009). Turbo roundabouts: Design principles and safety performance. *Transportation Research Record: Journal of the Transportation Research Board, Transportation Research Board, 2096*, 16–24. doi:10.3141/2096-03.
29. Fortuijn, L. G. H. (2009). Turbo roundabouts: Estimation of capacity. *Transportation Research Record: Journal of the Transportation Research Board, Transportation Research Board, 2130*, 83–92. doi:10.3141/2130-11.
30. CROW. (2008). *Turborotondes*. Publicatie 257, Dutch Information and Technology Platform.
31. Guerrieri, M., Corriere, F., Ticali, D. (2012, December). Turbo-roundabouts, a model to evaluate capacity, delays, queues and level of service. *European Journal of Scientific Research, 92*(2), 267–282, EuroJournals Publishing, Inc.
32. Tollazzi, T., Renčelj, M., & Turnšek, S. (2011). New type of roundabout: Roundabout with "depressed" lanes for right turning—"flower roundabout". *Promet—Traffic and Transportation, Scientific Journal on Traffic and Transportation Research, 23*(5), 353–358. doi:10.7307/ptt.v23i5.153.
33. Tollazzi, T., Jovanović, G., & Renčelj, M. (2013). New type of roundabout: Dual one-lane roundabouts on two levels with right-hand turning bypasses—"target roundabout". *Promet—Traffic and Transportation, Scientific Journal on Traffic and Transportation Research, 25*(5), Zagreb. doi: 10.7307/ptt.v25i5.1230.

Chapter 7
Non-motorized Participants on Alternative Types of Roundabouts

7.1 Introduction

In order to prevent accidents between motorized and non-motorized participants, the following different strategies are known in general:

- eliminate the risk;
- separate the non-motorized participants from the risk situation;
- if that is not possible—insulate the risk;
- if that is not possible—modify the risk;
- if that does not work—equip the road infrastructure (with additional road furniture), than control the risk behavior (watch, supervise);
- when it is not enough—inform and instruct non-motorized (also motorized) participants (through brochures, leaflets, newspapers, television…);
- when that cannot be done—restrict to approach the risk zone (by legislation or prohibition);
- the last action to be taken is to start the emergency (reconstruction and new—more safe solution).

All the entire above mentioned can be implemented also on roundabouts and especially on alternative types of roundabouts. As already known, the alternative types of roundabouts have some common features:

- lower number of conflict spots;
- acceptable in urban areas due to low dimensions (instead of standard "diamond", "diverging diamond", "cloverleaf" interchange);
- large capacities.

Due to the above-mentioned ("large capacity" could also means high speeds, and "acceptable in urban areas" usually means great number of non-motorized participants), we need to pay a lot of attention for managing non-motorized participants on

T. Tollazzi, *Alternative Types of Roundabouts*, Springer Tracts on Transportation and Traffic 6, DOI 10.1007/978-3-319-09084-9_7

alternative types of roundabouts. Otherwise, we could expect problems with children, blind, visually impaired, deaf, and physically impaired (elderly people) pedestrians.

The author of this book is not an expert in this field, but he believes that this chapter simply has to be included in the book. Mainly because we, the civil and traffic engineers, too often address only technical elements of our solutions, while forgetting about the needs and requirements of end-users. The author has no illusion as to describe in this book all the problems of impaired pedestrians, and he has not provided all suggestions for the solution of these problems. The main purpose of this chapter is just to raise awareness of civil and traffic engineers and experts that they have to consider also impaired pedestrians when designing technical solutions.

But let us start at the beginning. It needs to be stressed again that certain solutions that are suggested in one country could be dangerous in another. Consequently, most countries have their own guidelines for the geometric designing of roundabouts which are, as far as possible, adapted to real circumstances (local customs, habits, traffic culture, human behavior, tradition etc.) within these countries and are therefore the most acceptable within their surroundings. In cases of roundabouts, there is not "only one truth"; therefore, each country needs to "walk its own path". Verbatim, the copying of foreign results could be dangerous and could lead to consequences that are completely the opposite than expected.

At first, it is necessary to know that non-motorized participants' safety at a roundabout depends primarily on design elements of a roundabout, than on pedestrian and cyclists crossings and visibility, and slightly less on the traffic signs and road markings.

Equipping roundabouts with pedestrian crossings is necessary for providing safety and comfort of pedestrians, whereby we must avoid causing excessive traffic hold-ups. The availabilities of marked pedestrian and cyclists crossings at a roundabout are necessary in order to provide sufficient traffic safety and convenience for them, on condition that they do not cause excessive congestion of motorized traffic. A pedestrian and cyclists crossing will serve its purpose well only if it is located at a location where it attracts the greatest possible number of pedestrians and cyclists (who would otherwise cross the road randomly), and if it is sufficiently visible for drivers of motorized vehicles in order for them to stop their vehicle in time.

Pedestrian crossings should be implemented somewhat away from the roundabout exits, which results in a conflict between demands of pedestrians and drivers. If the pedestrian crossing is too far from the roundabout exit, pedestrians will not use it, whereas if it is too close, there is a possibility that the vehicles will accumulate all the way to the circulatory carriageway, hindering the circulatory traffic flow. In most cases, the recommended distance between the roundabout exit and pedestrian crossing amounts from one to three lengths of a passenger car [1]. Pedestrians and cyclists have to be able to perceive timely vehicles exiting or entering a roundabout. Special attention has to be paid to visibility for pedestrians in a roundabout combined with bus stops. Buses stopping at bus stops must not reduce the visibilities to pedestrians and/or drivers. The design of splitter islands has an impact on increased traffic safety for motorized and also other road users. Consequently, providing splitter islands is recommended even in cases where all the conditions are not met (e.g. sufficient width).

Traffic signs and road markings only influence traffic safety immediately after the roundabout is constructed, whilst later, their impact decreases or becomes void.

All the above mentioned is even more important for alternative types of roundabouts, and it doesn't depend on type, recent or at development phases.

7.2 Non-motorized Participants on Recent Alternative Types of Roundabouts

As pointed before several times, there is not only one truth in the case of roundabouts, and design solutions differ from country to country (differences between different countries have been attributed to different human behavior and duration on roundabout experience). However, the problem of traffic safety of non-motorized traffic participants on recent alternative types of roundabouts is tackled in different ways, particularly by:

- speed control at the design phase;
- separating island at pedestrian crossings;
- deviated position of the cycle crossing at the entry and exit;
- raised platforms at the crosswalks; and
- leading non-motorized participants at different level.

7.2.1 Speed Control at the Design Phase

It is common knowledge that the curvature of the driving curve (vehicle's path) through the roundabout has one of the greatest impacts on the traffic safety level. If the roundabout connections are implemented tangentially, the capacity is great, while the traffic safety level is quite low. For the driver tangentially connecting into the roundabout, it is hard to understand that he must give way to the vehicle within the roundabout, because he feels that he is on the priority road. Similarly also goes for the tangential exits, which puts the safety of pedestrians and cyclists, crossing the roundabout's arm, in danger. Even worse situation occurs, when the roundabout entries and exits tangent the central island, because in this case, the drivers can drive through the roundabout without reducing their speed. In these cases, the drivers do not have to reduce their speed. The driving curve (vehicle's path) in this case is almost straight.

At the selection of the roundabout dimensions (radii), the drive-through speed is one of the most important criteria. Lower speed of the motorized traffic leads to calmer traffic, whereas more attention can be devoted to other traffic participants and the possibility of serious traffic accidents is also reduced. The starting point, at the control method of driving through recent alternative types of roundabouts (e.g. turbo roundabout [2]) is that the drive-through speed does not exceed 30 km/h or 35 km/h. At these roundabouts the control of drive-through speed is required in three situations (Fig. 7.1):

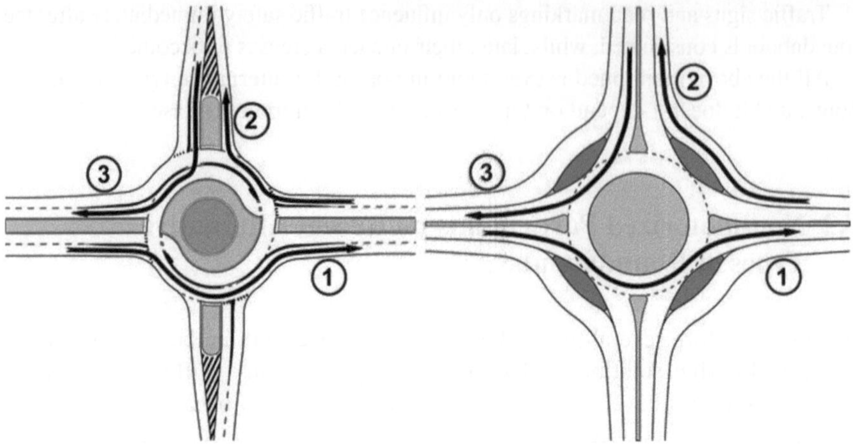

Fig. 7.1 Drive-through speed controls in three situations at turbo and flower roundabout

The first control (1) relates to the "traffic flow, driving through the roundabout". Depending of the type of alternative roundabout and number of arms we can create 1–6 driving lines for this directional traffic flow. The control is carried out separately, for each of the driving lines. Driving lines are composed of three curvatures of equal radii and of opposite direction [1].

The second drive-through speed control (2) is carried out for the "right-turning traffic flow from the right entry lane". Depending of the type of alternative roundabout and number of arms 2–4 driving lines can be created.

The third drive-through speed control (3) is carried out for the "right-turning traffic flow from the left lane of the side traffic direction" (usually in all recent alternative types of roundabouts). For this directional traffic flow, 2–4 driving lines are possible, depending of the type of alternative roundabout and number of arms.

The drive-through speed in each case is calculated, using the standard equation [2]:

$$V = 7,4 \cdot \sqrt{R_{outer}} \quad (1)$$

where V is in km/h and R_{outer} is in meters.

It needs to be stressed again that the drive-through speed must not exceed 30 km/h or 35 km/h.

7.2.2 Separating Island at Pedestrian Crossings

The correct implementation of pedestrian and cycle crossings is especially important at the recent alternative types of roundabouts, mostly because they have normally two-lane entries and/or exits and a pedestrian needs to cross approximately 7 m wide carriageway at a time (see Fig. 7.2). Because of such a long length of walk, a

Fig. 7.2 Pedestrian crossing two-lane entries and exits; Brno, Czech Republic

pedestrian is for a longer time in danger. In such cases, it is desirable to implement one of the measures to improve road safety for pedestrians and/or cyclists). One of such measures can be separation of entry and exit lanes with intermediate splitter islands—separating islands at pedestrian crossings (Figs. 7.2, 7.3, 7.4 and 7.5).

Separation of entry and exit lanes with an intermediate splitter island is implemented when a single traffic lane at the entry/exit is not acceptable in view of capacity and when sufficient space is available therefore.

Suggested width of an intermediate splitter island is 2.0 m (stroller + person who drives + protective width on either sides or bicycle + protective width on either sides), and minimum of 1.5 m [3].

7.2.3 Deviated Position of the Cycle Crossing

Deviated position of the cycle crossing at the entry and exit (Fig. 7.6) is popular in The Netherlands, but only out of urban areas, and where cyclists do not have a right-of-way.

This solution may only be used in extreme situations, when the flow of non-motorized participants is not strong, when more safe solutions cannot be applied and even in these cases only exceptionally and together with the implementation

Fig. 7.3 Separation of entry and exit lanes with separating islands at pedestrians' and cyclists crossing; Koper, Slovenia

Fig. 7.4 Separation of entry and exit lanes with separating islands at pedestrians' and cyclists crossing; Koper, Slovenia

Fig. 7.5 Separation of entry and exit lanes with separating islands at pedestrians' and cyclists crossings; Koper, Slovenia

Fig. 7.6 Deviated position of the cycle crossing at the entry and exit; near Den Haag, The Netherlands

Fig. 7.7 Design of deviated position of the crossing; near Den Haag, The Netherlands

of other measures for improving the traffic safety of non-motorized participants (e.g. optical or noise measures for traffic calming), when they cross the roundabout arm.

Deviated position of the crossing is designed in a way that it prevents high speeds of cyclists at the crossing, with deviated position of the cycle crossing on the area of the splitter island, for the width of a two-way cycle track and with deviation outward the roundabout mouth for approximately 10 m (see Fig. 7.7).

At these crossings of the arms by the deviated cycle track, the cyclists, on the splitter island, prior to crossing the arm, must be deprived of their priority.

Deviated position of the crossing is designed in a way that on both occasions (at the entrance and at the exit of the roundabout) cyclists at a splitter island is in a visual contact with motorized participants (looking at each other).

7.2.4 Raised Platforms at the Crosswalks

Raised platforms (sometimes called "flat top speed humps", "trapezoidal humps", "speed platforms", "raised crosswalks', "raised crossings", "speed table", "plateau", or "trapezium") are long raised speed humps with a flat section in the middle and ramps on the ends (Fig. 7.8); sometimes constructed with brick or other textured materials on the flat section.

Fig. 7.8 Raised platforms at roundabout's arm; Ljubljana, Slovenia

Their application is very useful on local and collector streets, and on main roads through small communities.

It is correct if they are long enough for the entire wheelbase of a passenger car to rest on the top [2].

This type of a measure for traffic calming is suggested by the Dutch guidelines for turbo-roundabouts [2] in a case of high speeds in front of a turbo-roundabouts (Fig. 7.9).

It needs to be pointed out that careful design is needed for drainage [4], and that this type of measure cause delay for fire trucks [5].

This type of a measure for traffic calming works well in a combination with textured crosswalks, curb extensions, and curb radius reductions, and as a rule include crosswalk and/or cyclists crossing. If crosswalk or/and cyclist passage is included, LED lights are added as a rule (Fig. 7.10).

7.2.5 Leading Non-motorized Traffic Participants at Different Levels

Pedestrian crossing should not compete with pedestrian overpasses or underpasses. Where there is no chance of providing sufficient visibility, where there is great intensity of the motor vehicles' traffic flow, great percentage of cargo vehicles in the traffic flow structure or a great number of pedestrians (and/or cyclists), level pedestrian crossings should not be implemented.

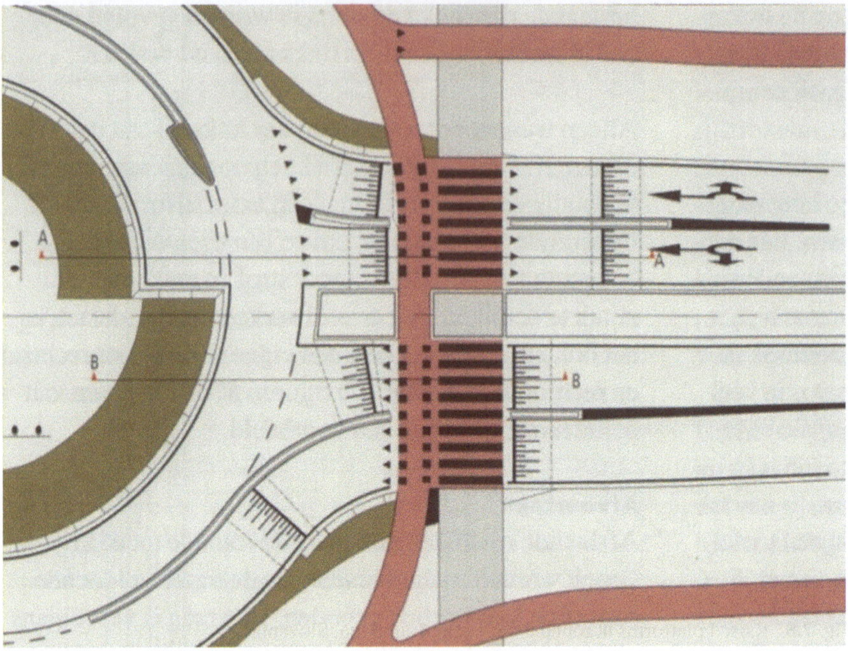

Fig. 7.9 Raised platforms by the Dutch guidelines for turbo-roundabouts [2]

Fig. 7.10 Raised platform with a crosswalk and LED lights in front of a mini-roundabout; Maribor, Slovenia

In roundabouts with two or more lanes at the entry (namely also in turbo-roundabout), it is not recommended to implement level pedestrian and cycle crossing in urban areas. In these cases, we must verify and substantiate whether there is reasonably constructing an underpass or overpass. In several European countries an underpass (Fig. 7.11) is usually a better solution than an overpass (Fig. 7.12).

Each case of the roundabout requires individual and thorough examination, whereas when taking a decision, we must consider the roundabout's size (number of lanes in the circulatory carriageway, speed of vehicles), intensity of pedestrian/cycle and motor-vehicle traffic flows (number of conflict situations) and number of lanes at entries/exits (length of the pedestrian crossing).

Indicative criteria, which must be met for the sensible and justified implementation of the overpass/underpass in some countries (also Slovenia), are the following:

- the product of the number of motor vehicles and the number of pedestrians in 24 h exceeds 150.000; or
- in the peak hour, 600 or more pedestrians crosses a roundabout's arm; or
- the share of cargo vehicles in the total traffic exceeds 12 %.

Fig. 7.11 An underpass at a turbo-roundabout on a main town arterial; Maribor, Slovenia

Fig. 7.12 An overpass at a turbo-roundabout on a main town arterial; Maribor, Slovenia

If at least one of the above-mentioned criteria is met, the implementation of the overpass/underpass is sensible and justified in the view of providing a sufficient level of the traffic safety.

7.3 Non-motorized Participants on Alternative Types of Roundabouts at Development Phases

Alternative types of roundabouts at development phases (flower, target, with segregated left-hand turning slip lanes…) are usually intended for rural areas, where normally we do not have non-motorized participants. If we plan to do such this type of roundabout in urban area, we could solve a problem of traffic safety of non-motorized participants in different ways.

Solutions will differ from country to country and it seems there will be no consistent rules. But, we can say that all measures at recent alternative types of roundabouts are in global appropriate also in cases of alternative types of roundabouts at development phases (speed control at the design phase, separating island at pedestrian crossings, deviated position of the cycle crossing at the entry and exit, raised platforms at the crosswalks, and leading non-motorized participants at different level). Anyway, only time will tell saying.

7.4 Impaired Pedestrians at Alternative Types of Roundabouts

Pedestrians at roundabouts cross during a gap in traffic flow, or when a vehicle has yielded for them, which may be quite difficult for pedestrians who are:

- young children: because their ability to determine speed and distance of approaching vehicles is not yet developed;
- visually impaired or blind: who cannot see a gap in traffic flow or detect yielding vehicles well and who might have difficulty determining the driver's intentions;
- hearing impaired or deaf: who cannot hear approaching vehicles;
- elderly: due to deteriorated ability to determine gaps or recognize the speed of approaching vehicles, and their slower walking speed;
- physically impaired: who cannot walk at a normal speed, or may be using a wheelchair and be at a lower height for viewing vehicles and may not be able to see beyond stopped vehicles or see the driver of stopped vehicles;
- people with cognitive disabilities: who need clear-cut rules.

These type of pedestrians need the safety benefits that roundabouts provide as much as (or more than) others. Roundabouts can and should also be accessible for these types of users.

The problem of traffic safety of impaired pedestrians at alternative types of roundabouts is tackled in different ways, particularly by measures mentioned in Sect. 7.2. But, there are also countries with special, less widely used measures, as signalized crosswalks [6–8], barriers or distinct elements to prevent blind persons from inadvertently crossing a roundabout roadway in unsafe locations. Tactile ground surface indicators, TGSIs [9], or detectable warning surfaces [10] (Figs. 7.13 and 7.14), with a distinctive bumpy texture under foot, are used to help people who are blind or who have low vision recognize the edge of the street and the edges of the splitter island [11].

At the end it needs to be stressed that the most research on the field of impaired pedestrians at roundabouts has been made in the USA, where the Americans with disabilities act requires consideration of these types of pedestrians in building new intersections and facilities.

Fig. 7.13 Tactile ground surface indicators or detectable warning surfaces are installed along the curb and splitter island at a pedestrian crossing; Amsterdam, The Netherlands

Fig. 7.14 Tactile ground surface indicators; Raleigh, North Carolina, USA

References

1. Ministrstvo za promet. (2011). Tehnična specifikacija za javne ceste, TSC 03.341: 2011 Krožna križišča (Slovenian Guidelines for Roundabout Design), Ljubljana, Slovenia.
2. CROW. (2008). Turborotondes, Publicatie 257. The Netherlands: Dutch Information and Technology Platform.
3. *Pravilnik o cestnih priključkih na javne ceste* (2009). Uradni list RS, št. 86/2009, Ljubljana.
4. CROW. (1998). ASVV: Recommendations for Traffic Provisions in Built-up Areas. The Netherlands: Report 15.
5. Institute of Transportation Engineers. (2014). Traffic calming measures—Speed table. Washington, USA. http://www.ite.org/traffic/table.asp. Accessed April 10 2014.
6. Baranowski, B. (2014). Pedestrian crosswalk signals at roundabouts: Where are they applicable? ITE District 6 Annual Meeting, Sacramento, California, USA, 20–23 June 2004, http://www.k-state.edu/roundabouts/research/bill2004.pdf. Accessed April 2 2014.
7. Schroeder, B., Hughes, R., Rouphail, N., Cunningham, C., Salamati, K., Long, R., et al. (2011). Crossing solutions at roundabouts and channelized turn lanes for pedestrians with vision disabilities, NCHRP Report 674, Washington, D.C.: TRB.
8. Schroeder, B., Hughes, R., Rouphail, N., Cunningham, C., Salamati, K., Long, R., et al. (2010). Supporting material to NCHRP Report 674, NCHRP web only document 160, Appendices B-N to Contractor's final Report for NCHRP project 3-78A. Washington, D.C.: TRB.
9. RTS 14. (2007). Guidelines for facilities for blind and vision-impaired pedestrians, Australia and New Zealand Standard Organization (2nd ed.). http://www.nzta.govt.nz/resources/road-traffic-standards/docs/draft-rts-14-revision-2007.pdf. Accessed May 16 2014.
10. US Access Board. (2014). Proposed accessibility guidelines for pedestrian facilities in the public rights-of-way, Section 305, Released 26 July 2011, http://www.access-board.gov/guidelines-and-standards/streets-sidewalks/public-rights-of-way/proposed-rights-of-way-guidelines/chapter-r3-technical-requirements. Accessed May 16 2014.
11. Kenjić, Z. (2009). Kružne raskrsnice-Rotori, Priručnik za planiranje i projektovanje. IPSA, Sarajevo.

Chapter 8
Possible Ways of Roundabouts' Development

8.1 Present Position

Today, modern roundabouts exist in all European countries, and there are also several countries elsewhere in the world where they are numerous (the US, Australia, New Zealand, Israel, and Mexico). We can say that modern roundabouts are a world phenomenon today.

No uniform guidelines exist in Europe for the geometric design of roundabouts as specific circumstances differ from country to country. Certain solutions that are safe in one country could be less safe in another. Consequently, most of countries have their own guidelines for the geometric design of roundabouts, which are adjusted to their actual conditions (general culture, traffic culture, local custom…) and are thus the more acceptable within their environment.

Roundabouts in different countries also differ in their dimensions and designs; the reasons being the different maximum dimensions of motor vehicles (mostly heavy vehicles) and specific human behavior. As has been pointed out several times there is not "only one truth" in the case of roundabouts. Each country must find its own way. This is the hardest and the slowest but at the same time also the safest way. Any word-for-word copying of other country's solutions can be dangerous and can cause completely opposite effects than expected.

In Europe (the same applies for the rest of the world), different countries are at different stages of a development concerning roundabouts. If we disregard the UK (which has always traditionally been on the top), over the past two decades there has been intensive roundabouts' development in some other European countries (Austria, Slovenia, Croatia, Czech Republic, the Former Yugoslav Republic of Macedonia) besides France, The Netherlands and Germany. At the moment there is intensive roundabouts' development in Italy as well. As mentioned above (different countries at different stages of development), means nothing wrong, as roundabouts within each country are developed according to their intensities, regarding what is acceptable for their environments and end-users.

T. Tollazzi, *Alternative Types of Roundabouts*, Springer Tracts on Transportation and Traffic 6, DOI 10.1007/978-3-319-09084-9_8

It needs to be stressed that the roundabout intersection has been "at the development phase" since 1902, and this development is still in progress. In several European countries (The Netherlands, Germany, France, Switzerland, as also Slovenia), research into the various aspects and the various types of roundabouts as a useful type of road intersection has spanned many decades. During this long period the number of vehicles, their sizes and speeds and in particular their acceleration capabilities have radically changed. The same situation applies to drivers' experiences and their expectations on the highway infrastructure. In addition, with increased traffic there may be more concern about matters of public safety and liability. These changes have had a strong influence on the evolution of modern roundabouts especially over the last two decades, and a little more.

There is now a good level of understanding of the requirements for capacity and safety at roundabouts. The major elements of predictive design geometry have been dealt by research into empirical formulae, micro simulations, and by some case studies (pilot projects) in real life.

8.2 Possible Directions of Development

Roundabouts' development in the future is difficult to predict but it is possible to state that:

- development of new types of roundabouts will also continue in the future;
- the number of roundabouts will still be increasing;
- in Europe more examples of alternative roundabout types will also be implemented;
- we can expect the highest growth in the number of modern roundabouts and also alternative types of roundabouts especially in the US, Canada, Japan, and in some other countries.

We can presume that in the near future mostly one-lane roundabouts (mini, small and medium) and alternative types of roundabouts will be implemented. The former, because they represent a good measure for traffic calming in urban areas, and the latter, because they are intended for solving specific traffic situations.

8.3 Some Areas for Future Researches

As primary influences of modern roundabout's layouts in relation to capacity and traffic safety are now known, there may be an opportunity to gain better understanding of the effect of secondary factors. This would provide a clearer view in some areas where there are doubtful or even conflicting notions of good practice. There are some other areas, for example:

- new alternative types of roundabouts;
- capacities of alternative types of roundabouts;
- traffic signal controlled roundabouts;

- the influences of roundabouts on environment;
- provisions for cyclists;
- elderly drivers and pedestrians on roundabouts;
- blind pedestrians at roundabouts;

For which the safety, environmental and economic justification for extensive research into rather intractable problems might need to be made.

As it is now known, alternative types of roundabouts typically differ from "standard" one- or two-lane roundabouts in one or more design elements, as their purposes for implementation are also specific. Main reasons for their implementation are particular disadvantages of "standard" one- or two-lane roundabouts regarding actual specific circumstances. Usually, these disadvantages are highlighted by low-levels of traffic safety or capacities. As already known, an one-lane four-arm roundabout has theoretically only 8 conflict spots (4 merging and 4 diverging). If there are two circular lanes, the number of conflict spots increases by the number of weaving conflict points, which theoretically equals the number of arms. But, from the practical point of view, we are not only speaking of conflict spots at the multi-lane roundabouts, but also of conflict sections (sequence of conflict spots), since there is no predetermined spot for drivers where they must change lanes along the circulatory carriageway. At multi-lane roundabouts with two-lane entries and exits, the traffic-safety conditions are even slightly worse. In this case, there are conflicts at the points of crossing the circulating lanes at the entries and even greater ones in the course of changing traffic lanes along the circulatory carriage ways. However, by far the most dangerous is the maneuvering when leaving the roundabout. Lately, a growing number of foreign analyses have pointed to the poor traffic-safety characteristics of "standard" multi-lane roundabouts and poor experiences related thereto. Therefore, many countries are looking for solutions as what to do at these existing roundabouts in order to improve the higher level of traffic safety. Different countries tackle this problem in different ways, which can be divided into four groups. A higher level of traffic safety at "standard" multi-lane roundabouts can be achieved by:

- decreasing the number of driving lanes in the circulatory carriageway: not a good solution because the roundabout's capacity is decreased;
- decreasing the number of driving lanes at entries/exits: not a good solution because the roundabout's capacity is decreased;
- increasing the outer roundabout's diameter (whereby, the available length for weaving within the circulatory carriageway is increased): financially very demanding;
- decreasing the number of conflict spots: a good compromise between the finances on the one hand and the increased capacity and traffic safety level on the other.

Recently, several countries are solving the problem of low traffic safety of "standard" two-lane roundabouts by adopting the last of the above-mentioned methods; by decreasing the number of conflict spots. One way to decrease the number of conflict spots are alternative types of roundabouts (such as turbo-roundabouts, flower roundabouts, and some other types). One of the main characteristic of alternative types of roundabouts is that there are no weaving conflict spots.

All alternative types of roundabouts have their advantages and deficiencies, which makes sense since they are intended to solve particular problems. In the near future, we can expect further developments regarding alternative types of roundabouts, with the intention of solving specific problems.

Traffic simulation is being increasingly used to assess traffic operations along many different types of roadway networks, and several proprietary microscopic simulation programs are available for the modelling of general traffic networks at the moment. Recent alternative types of roundabouts are relatively new; consequently, practical evaluation data are presently unavailable for them. For example, only in The Netherlands a number of turbo-roundabouts have been constructed and very few of those are operating at/or near capacity.

A slightly different situation is in a case of alternative types of roundabouts at development phases because they do not as yet exist, and it is impossible to presume and calibrate any of the influenced factors. As alternative types of roundabouts at development phases are practically unique, there are no specific program tools for them. Accordingly, software tools that allow the designer to create unique solutions are an advantage. Therefore, different possibilities remain open for determining the capacities of alternative types of roundabouts at development phases.

Traffic signal controlled roundabouts have also been at the development stages, especially in last two decades. Introducing traffic signal control to a roundabout is a technique that can be used to overcome problems associated with uneven traffic flows during peak periods. As is known, at the moment four options exist for introducing traffic signal controls to roundabout; direct, indirect, full, and part-time.

Although the traffic signalization of roundabouts is discouraged in the US, full signalization of the circulatory roadway is widely used in Europe (France, The Netherlands, Sweden, Poland) as also in the UK and Australia, and has also started to become very popular elsewhere.

The majority of published literature on the subject of traffic signal controlled roundabouts is based on the experiences in the UK, where they were adopted initially to control traffic at many of the older large roundabouts that have existed for many years. The use of full time or total signal control at these roundabouts has been the method used in the majority of their cities.

In Australia, the practice has been to install signals on one approach to the roundabout to meter circulating flow in order to reduce delays on the opposing arms of the roundabout. This approach operates on a part-time basis and is activated by the queue length on the delayed approach reaching a defined critical distance.

Good experiences with traffic signal controlled roundabouts are also reported from Poland and Sweden. In Poland, traffic signal controlled turbo-roundabouts represent a new generation of junctions that offer many advantages for urban traffic such as higher capacity, increased safety, and shorter pedestrian crossings.

Therefore, in several countries it is appropriate or even necessary to implement traffic signals:

- at roundabouts where traffic conditions changed after their implementation;
- at existing, traffic overloaded roundabouts;

- where a tram line or a suburban railway intersect a roundabout;
- to increase traffic safety regarding heavy volumes of pedestrians and cyclists.

From all the foregoing, it is evident that in this area there are still several opportunities for research.

The environmental criterion has started to be as important as safety criteria. We can say that in some countries this criterion became the most important criteria today; especially in urban areas where we are looking for answers as to which type of intersection would have minimum negative impacts on the environment (on the quality of life). Road pollutant emissions, above all within the urban context, are correlated to several infrastructural parameters and to traffic intensity and typology. Roundabouts as a rule provide environmental benefits by reducing vehicles' delays and the numbers and durations of stops, compared with signalized or all-way stop-controlled intersections. On roundabouts, even when there are heavy volumes, vehicles continue to advance in slowly moving queues rather than coming to a complete halt. This reduces air quality impacts and fuel consumption significantly by reducing the number of acceleration/deceleration cycles and the time spent idling. In this area there are also many opportunities for research in terms of CO, CO_2, CH_4, NO, $PM_{2,5}$ and PM_{10} vehicular emissions, due to the software available for the calculations of emissions from the road transport sector.

Although in the field of cyclists' management a lot of novelties have already been introduced over recent years, we can expect new cyclists'-friendly and safety solutions; especially in countries with high numbers of cyclists.

Several studies have shown that an average age of the population in general is increasing. There are several reasons for that; the most important being a research in medical sciences, higher living standard in general, and also improvements in the qualities of micro-environments at work and at home. Statistical analyses show that people over 70 in the USA represent 9 % of the population, and in the European Union already 17 % of citizens are older than 65 years. It is a fact that most of these persons are still active participants in traffic, especially in city traffic. Therefore, in several countries, drivers over 60 years of age represent a rapidly growing part of the driving population.

It is well-known that the qualities of sight and hearing as well as flexibility and reaction times deteriorate with advancing age. Older drivers are more often considered responsible for crashes and they have more fatal accidents, especially because they are more vulnerable. Many different studies have shown that elderly drivers are more frequently involved in specific types of accidents (e.g. situations involving more than one vehicle, especially at intersections). Older drivers have more error accidents and this tendency increases with age. An error accident is defined as the failure of the planned action to achieve a desired outcome without the intervention of some chance or unforeseeable event. In general, these studies consistently find correlation between crash rates and older driver's traffic safety: accidents are more likely to occur in good weather, during daylight hours, at intersections, while making turns; their causes are not excessive speed or alcohol. These results show that the problems of older drivers in road traffic regarding overall safety is not negligible even today, while its importance will rise in the future.

In general, in urban areas there are many factors that contribute to the low level of traffic safety. One of them is certainly the higher number of intersections where drivers change their driving directions. This creates several trouble spots. Another reason is high the amount of information that drivers should pay attention to, the high number of signals allowing and forbidding something, and ultimately there are too many vehicles of different types concentrated within a small area, among them also delivery vehicles and trucks.

Today, many studies concentrate on the safety evaluation regarding elderly drivers at various types of intersections, in regard their own perspectives. The results show that recent types of roundabouts are too complicated for older road users, so we will need to pay more attention to this fact in future. Traffic experts will have to accept the fact that the percentage of the elderly population involved in traffic is increasing all the time. When planning traffic solutions it is very important for elderly drivers that the road infrastructure serves its function and that an user can predict how other drivers will react, especially at large multi-lane intersections. A driver needs enough time to make decisions and to act accordingly to rules and anticipations, and solutions should be made that would lower the risks for elderly traffic participants.

As crossing at a roundabout requires pedestrians to visually select a safe gap between cars that may not stop, or to cross in front of vehicles that have yielded to them, accessibility has been problematic. While roundabouts may be an asset to traffic planners in controlling and slowing the flow of traffic at intersections without using traffic signals, the lack of predictable opportunities to cross presents a problem for pedestrians with vision impairments, for pedestrians who are unable to move quickly, or for those who are uncertain or very cautious. Vehicle drivers may yield to pedestrians for a very short time, particularly at roundabout exits, moving on before the uncertain pedestrian has decided that it is safe to cross. Blind and visually impaired pedestrians rely primarily on auditory information to make judgments about when it is appropriate to begin crossing a street and the sound of cars circulating in the roundabout may make it difficult to hear a car approaching the crosswalk, or a car that has yielded to allow them to cross. The usefulness of non-visual information for crossing at roundabouts, and methods for making roundabouts more accessible, are under study, but many issues remain.

The benefits of maximum success from all the expected research could possibly be estimated, but it should be obvious that an expectation of future benefits from a particular item of research, as sometimes implied in funding requirements, cannot be guaranteed.

Disappointing results must sometimes be expected and negative results also have value!

MIX
Papier aus verantwortungsvollen Quellen
Paper from responsible sources
FSC® C105338

If you have any concerns about our products,
you can contact us on
ProductSafety@springernature.com

In case Publisher is established outside the EU,
the EU authorized representative is:
Springer Nature Customer Service Center GmbH
Europaplatz 3, 69115 Heidelberg, Germany

Printed by Libri Plureos GmbH
in Hamburg, Germany